Developing Secure Distributed Systems with CORBA

For quite a long time, computer security was a rather narrow field of study that was populated mainly by theoretical computer scientists, electrical engineers, and applied mathematicians. With the proliferation of open systems in general, and of the Internet and the World Wide Web (WWW) in particular, this situation has changed fundamentally. Today, computer and network practitioners are equally interested in computer security, since they require technologies and solutions that can be used to secure applications related to electronic commerce. Against this background, the field of computer security has become very broad and includes many topics of interest. The aim of this series is to publish state-of-the-art, high standard technical books on topics related to computer security. Further information about the series can be found on the WWW at the following URL:

http://www.esecurity.ch/seriseseditor.html

Also, if you'd like to contribute to the series by writing a book about a topic related to computer security, feel free to contact either the Commissioning Editor or the Series Editor at Artech House.

Recent Titles in the Artech House
Computer Security Series

Rolf Oppliger, Series Editor

For a listing of recent titles in the *Artech House Computing Library,*
turn to the back of this book.

Developing Secure Distributed Systems with CORBA

Ulrich Lang
Rudolf Schreiner

Artech House
Boston • London
www.artechhouse.com

Library of Congress Cataloging-in-Publication Data
Lang, Ulrich.
    Developing secure distributed systems with CORBA / Ulrich Lang,
        Rudolf Schreiner.
        p.   cm. — (Artech House computer security series)
    Includes bibliographical references and index.
    ISBN 1-58053-295-0 (alk. paper)
        1. CORBA (Computer architecture)   2. Distributed operating systems (Computers)
    3. Computer security.   I. Schreiner, Rudolf.   II. Title   III. Series.
    QA76.9.A73 L36   2002
    005.2'76—dc21   2001056566

British Library Cataloguing in Publication Data
Lang, Ulrich
    Developing secure distributed systems with CORBA. —
    (Artech House computer security series)
    1. Computer security   2. CORBA (Computer architecture)
    3. Distributed databases—Security measures
    I. Title   II. Schreiner, Rudolf
    005.8

    ISBN 1-58053-295-0

Cover design by Igor Valdman

© 2002 ARTECH HOUSE, INC.
685 Canton Street
Norwood, MA 02062

International Standard Book Number: 1-58053-295-0
Library of Congress Catalog Card Number: 2001056566

10 9 8 7 6 5 4 3 2 1

Contents

Foreword

Security has been elevated to a prominent position in today's IT world, ranking high in many studies investigating the issues that need to be addressed in the development of IT, the Internet, and e-commerce. Security is easy to do on paper, be it compiling lists of security requirements or proposing abstract security architectures. The devil, as so often, lies in the details, and the technical challenge in implementing security has been aptly described as programming Satan's computer [Anderson & Needham, Springer LNCS 1000]. It should therefore be no surprise that the literature on security is much stronger on abstract designs and on "solutions" that are not closely matched to actual problems than on the construction of concrete security systems, which is a slow and arduous process.

The history of CORBA security provides a fitting illustration of the points just made. The CORBA security architecture started off as a paper design with an impressive list of desiderata. Attempts to implement this architecture exposed various inconsistencies and areas that needed further work. Over the years, the CORBA security architecture has now developed in concert with its implementations. The demand to deliver CORBA services over the Internet has had a strong influence on this process. To some extent it has refocused the CORBA security goals by putting a stronger emphasis on communications security and on the interaction of CORBA security services with standard Internet security components like firewalls or the SSL/TLS protocol. CORBA security will keep evolving in reaction to the demands users are putting on the CORBA platform.

For real security, security has to be done for real. It is therefore particularly gratifying to see a book that is founded on a concrete implementation of a security system. This book presents the general goals and features of CORBA security in the context of the MICOSec system the authors have been developing. The advantages of this approach should be twofold. First, the views of the authors have been tempered by their practical experiences in trying to fit security mechanisms into the CORBA middleware framework. Secondly, the book can serve as a user guide to an actual CORBA security system. Thus, readers interested in evaluating the merits of CORBA security, or of middleware security in general, need not restrict themselves to a mere paper analysis but have a basis for experimentation to gain real experience. Real experience with CORBA security in a variety of applications is, in turn, the input that should drive the continuing development of the CORBA security architecture.

Thus, the book not only presents a snapshot of the current state of CORBA security, it hopefully can also help to turn the wheels of progress in this field.

Dieter Gollmann
Cambridge, U.K.
January 2002

Preface

In recent years, demand has increased for distributed applications in complex and heterogeneous IT systems (e.g., for telecommunications and banking). As part of the development of such applications, implementers had to repeatedly solve the same problems related to distributed computing. These problems ranged from technical details such as byte ordering and addressing servers in networks to processing complex transactions. To take this work off application developers, an abstraction layer can be introduced to segregate applications from the underlying communication layers. From the application perspective, this so-called middleware layer transparently takes care of all communications tasks. The most advanced middleware architecture today is the Object Management Architecture (OMA), which has been developed by the Object Management Group (OMG) consortium since 1989. The core component of the OMA is the Common Object Request Broker Architecture (CORBA), which defines an Object Request Broker (ORB) that provides the abstraction layer and automatically takes care of all underlying communications tasks. CORBA is mechanism independent (i.e., supports remote method invocations across several network types, across differing hardware platforms, and across differing programming languages).

In the first half of the 1990s, lack of effective security features prevented the deployment of many CORBA-based applications for the Internet or other insecure networks, such as e-commerce or network management applications. In response to the rising demand for security, the OMG specified the CORBA security services, which define a comprehensive set of security features for CORBA systems, ranging from authentication and access control to audit and nonrepudiation.

This book should be considered both as a textbook that explains the CORBA security services architecture and its design rationale in a precise and comprehensive manner and as a hands-on reference on how to use the CORBA security services in practice.

The first part of this book (Chapters 1–3) covers the architecture and design rationale behind the CORBA security services in detail. Chapter 1 introduces the main design features and essential use of CORBA to set the scene for the remainder of the book. Chapter 2 defines a number of basic security concepts in the context of CORBA and is intended for an audience that is not too familiar with information security terminology and theory. Chapter 3 comprises an in-depth, but at the same time conceptual, description of the CORBA security services architecture.

The second part of this book (Chapters 4–7) is concerned with the hands-on use of MICOSec, an Open Source implementation of the CORBA security services specification. Chapter 4 illustrates how MICOSec is installed and configured. Chapters 5 and 6 describe how you can use the application-facing interfaces that CORBA security offers from within your own applications. Chapter 7 then shows how CORBA security can be used to secure applications that should remain unmodified (e.g., legacy applications).

Acknowledgments

The authors would like to acknowledge the following people for their support and assistance during the writing of this book: Dima Skalmovsky and Svetlana Bagiyan, Ameneh Alireza, Regine Schickentanz and Gerald Lorang, Dieter Gollmann, David Chizmadia, Richard Soley, the MICO developer team, ObjectSecurity Ltd., and Artech House Publishers.

1

Introduction to CORBA

1.1 Why CORBA?

1.1.1 The Business Perspective

The first question many people might ask when they first hear about the Common Object Request Broker Architecture (CORBA) is why they would ever want to choose CORBA. This question is often asked by people outside the software engineering profession. Some of CORBA's more technical advantages are sometimes not easy to understand without knowledge of object-oriented programming or of the inherent communications difficulties in distributed systems. But even without a technical background, it is easy to see why CORBA is useful for distributed application development.

The main reason CORBA is important is because businesses today live in a world connected by computer networks, in which users need to share information across enterprises as never before. This information to be shared comes from many different sources, such as stand-alone applications on various differing hardware and software platforms, and only a few of them are designed to interoperate with other applications on their own platforms, let alone with applications on differing platforms. If an application does share information with other applications, it can normally only interoperate with a select few.

Integrating these isolated applications and systems is generally not easy. Converters between data formats and communications protocols are often necessary. This requires a custom solution, which can be time-consuming

1

and expensive because of the number of different applications that need to be connected. It is impossible to reuse software components that have been written for incompatible platforms. If updates to the system are necessary at a later point in time, even more time and money will be spent. In addition, components normally do not interoperate efficiently within ad hoc or proprietary integration solutions since the applications were not designed with interoperability in mind. Poor integration leads to organizational inefficiencies, such as redundant data entry and multistep data conversion. These processes are costly, time consuming, and error-prone. From a user perspective, poor integration results in the need to learn disparate applications, since user skills cannot be transferred from one system to another.

For example, managers need to be able to pull data they need from several sources (e.g., marketing department's graphics, finance department's spreadsheets, design data) over the network, including both current and historical data, to reach their business decisions. Or consider large companies, which often have systems that have been evolving over decades, including mainframes, personal computers, local-area networks (LAN), wide-area networks (WAN), and multiple database management systems. These systems run a diverse set of applications on different operating systems. Frequent changes in the software environment, due to new business needs, new technology, and organizational changes, could be carried out more efficiently in an integrated environment.

Another example is the management of different legacy systems in telecommunications environments, such as phone switching and signaling systems. A centralized, integrated management infrastructure could reduce administrational efforts, as well as the maximum accepted latency of the system.

1.1.2   The Technical Perspective

In a nutshell, CORBA is a specification for a software library the Object Request Broker (ORB) with standardized object interfaces that allow software objects to talk to each other across a network in a well-defined way. In addition, CORBA automatically applies a range of useful services to communications. CORBA's beauty is that it does all this largely transparently to the application programmer. After the ORB is initialized, all CORBA objects can be invoked like normal software objects.

The generic term for systems like CORBA is *middleware*—software that resides between an application and the inner workings of the system hosting the application. Middleware insulates applications from the software's lower-level details and complexities so that the application developer only has to deal with a single application programmers interface (API) of some sort—the middleware handles other details such as mediating communications to remote objects. Instead of coding to operating systems or low-level interfaces, the application developer can use middleware to work on a higher level in the application, while the middleware provides the lower-level details. CORBA is communications middleware because it insulates an application from the details of the communications kernel. However, software developed in accordance with the CORBA specification does not necessarily need to communicate over a network—it could just as well use methods that are contained in the client's own address space.

From a technical perspective, CORBA has many advantages, both for software engineers and the enterprise as a whole. The CORBA architecture exhibits the following general design features:

Transparency

CORBA hides many of the inherent difficulties of distributed object computing from the application programmer. All invocations of methods in remote objects will be handled transparently by CORBA. To the application programmer, all object calls appear to be local invocations. The application programmer does not even need to know where the object is located on the network at the point the invocation is carried out.

In addition, CORBA automatically provides a number of useful services to network communications, such as transaction processing or naming. Most of these services can be bought as add-on software packages, which can be plugged into the ORB to add the required functionality to CORBA. The main focus of this book, the CORBA security service, is one such service.

The fact that CORBA is largely transparent on the application layer simplifies the programming of applications in distributed systems and, therefore, can reduce the overall application development cost, as application programmers do not have to be CORBA specialists to use CORBA to connect their application components.

Platform independence

CORBA has its own language to describe object interfaces, the so-called Interface Definition Language (IDL), which can be compiled into a variety of target programming languages and platforms. This way, CORBA interfaces are independent of the programming language used to implement the client and server objects. For example, it is possible to have CORBA interfaces for a browser-based Java front end, a PC-based C++ middle tier, and a

UNIX-based C back end. All these software objects can be connected across this heterogeneous environment with CORBA because all object interfaces have an IDL representation.

CORBA also provides its own communications protocols, which run on top of a variety of conventional network protocols (e.g., TCP/IP).

Portability

A CORBA application can be easily ported from one ORB product to another—as long as there is an ORB product that support the same programming language for which the CORBA application object was originally developed, of course. Porting of applications is possible because CORBA standardizes the interfaces visible from the application layer, so all compliant ORBs exhibit the same interfaces to which the application is connected. In practice, however, there are minor differences between ORB products from different vendors, so a little bit of tweaking may be necessary to port applications between differing ORBs.

In theory, CORBA's add-on ORB level services, such as security and transaction processing should also be portable because CORBA standardizes interfaces between the ORB and these services, which all compliant CORBA vendors have to support. These interfaces, appropriately called *interceptors*, intercept messages as they pass through the ORB and apply service-specific functionality, such as encryption or transaction processing, to them.

Software reuse

CORBA is best at avoiding unnecessary development costs if many software objects can be reused in various parts of the system, because CORBA provides the means to access all objects flexibly across platform and programming language boundaries. This means that software does not need to be ported to different programming languages or platforms anymore, hence the total cost and effort of software development and maintenance can be reduced.

Integration

As mentioned earlier, CORBA is platform- and programming-language–independent, and therefore allows the integration of software components from various sources. For example, a Java front end in a Web browser can use information from a UNIX-based back-end customer database and from an AS/400-based accounts database to provide an integrated view of customers and their accounts. Most major enterprises have a large number of legacy systems that contain information that is critical to the business but that can often not automatically be tapped into from modern systems because of incompatibilities of data formats and communication protocols. In practice, CORBA's solution to this problem is often its biggest selling point, and many companies successfully managed to develop so-called "CORBA wrappers" to provide an IDL interface to their legacy systems.

CORBA achieves integration across networks with differing technologies through its own communications protocols that can run on a variety of underlying transport mechanisms. These CORBA protocols standardize message and data formats so that the nature of the underlying network protocols and topology becomes irrelevant as long as the transport mechanism can convey CORBA messages. The most commonly used CORBA protocol is the Internet-Inter-ORB Protocol (IIOP), which specifies how CORBA messages are transported over the Internet via TCP/IP.

Interoperability

The core idea behind CORBA is that interoperability between objects running on compliant CORBA products from different vendors is possible because CORBA specifies its own standardized communications protocols and interface definition language. Therefore—at least in theory—all compliant CORBA products should be able to interoperate.

However, it is important to understand that CORBA cannot always provide interoperability if the underlying technology does not match. For example, if two different security service implementations use different cryptographic algorithms to protect their communications, then CORBA will not be able to abstract from these inherent incompatibilities. How should the recipient know how to decrypt a message encrypted with an unknown or unsupported cryptographic algorithm?

Flexibility

CORBA is largely mechanism-independent, which means that ORBs will be able to integrate objects from any platform as long as a valid language mapping exists, and from any underlying technology as long as the mechanisms can be fitted to the CORBA interfaces. For example, CORBA's communications protocols can be mapped onto a variety of different transport mechanisms. CORBA add-on object services are also designed to support a variety of underlying mechanisms, such as different cryptographic algorithms in the security service.

This feature of CORBA makes it possible to use CORBA in a variety of application domains. For example, both telecommunications providers and medical health care companies have successfully deployed CORBA. From a security perspective, CORBA has been used to secure a wide range of different application types (e.g., just-in-time communications at large manufacturing companies, as well as Internet e-commerce applications).

Location Transparency

CORBA today needs to be able to support mobile applications, for example, in wireless environments with mobile devices where servers change their location often. Such objects are often called nomadic, and CORBA's naming and trading services, as well as its location-transparent invocation mechanism, provide the means to locate and invoke these nomadic objects.

Scalability

CORBA was always intended to support environments with a potentially large number of objects and users. Therefore, the CORBA architecture was designed in such a way that it would not pose any restrictions on the number of objects and users in the system. It also tries to actively support the management of large systems.

1.2   The Object Management Group

CORBA was first published in 1990 by the Object Management Group (OMG) [1], a nonprofit organization founded in 1989, that has since grown to more than 800 member companies, representing the entire spectrum of the computer industry. The OMG is the world's largest software consortium, and its objective is to establish object-oriented industry guidelines for integrating distributed applications based on a variety of existing technologies. To achieve this, OMG specifications are designed to be both rich enough to be useful as a standard and flexible enough to accommodate a wide variety of distributed systems.

The first key specifications adopted by the OMG is the Object Management Architecture (OMA), which provides a complete architectural framework for CORBA that is rich and flexible at the same time. The definitions of CORBA's main parts are contained in the *Common Object Request Broker Architecture* [2] and *CORBAservices: Common Object Services Specifications* [3]. A copy of the specifications and information on updates can be obtained from the OMG Web site, http://www.omg.org.

The OMG standardization process is purely vendor-driven and open to all members, and all submissions in reply to request for information (RFI) and request for proposal (RFP) documents have to be implemented by the submitting company within a specified timeframe. The final standard is adopted by a voting scheme based on both business and technical merit. There are a large number of separate committees in the OMG for the CORBA platform, CORBA services, and application domains.

1.3 The OMA

1.3.1 Introduction

In November 1990, the OMG defined an object-oriented architecture for platform-independent application interactions called the OMA [4], which is an abstract umbrella architecture for all OMG specifications. It categorizes the areas of CORBA standards depending on purpose and level of abstraction.

The OMA is composed of an *object model* and a *reference model*. The object model defines how objects distributed across heterogeneous environments can be described, while the reference model characterizes interactions between those objects. Through adherence to the OMA, CORBA enables the development and deployment of interoperable distributed object systems in heterogeneous environments.

In the OMA object model, an object is an encapsulated entity with a distinct immutable identity whose services can be accessed only through well-defined interfaces. Clients issue requests to objects that perform services on their behalf. The implementation and location of each object is hidden from the requesting client. Although the OMA implies that all participating software components are objects, it is important to note that applications need only support or use OMG-compliant interfaces to participate in the OMA. They need not themselves be constructed using the object-oriented paradigm. Existing nonobject-oriented software can be embedded in objects called object wrappers that participate in the OMA.

The OMG OMA reference model groups object interfaces into interface categories that are conceptually linked by an ORB. At the bottom of the standards is the ORB that implements the communication infrastructure through which all CORBA-compliant objects communicate. The ORB mediates all communications between objects and transparently activates those objects that are not running when they are invoked. The CORBA services are the fundamental interfaces bridging the gap between clients and the ORB, and they are often compared with operating system service calls. Located on top of the ORB and CORBA services are the CORBA domains and CORBA facilities. The CORBA domains are vertical facilities like financial services and health care, whereas CORBA facilities are standards and services used horizontally across applications and domains (e.g., system management).

Figure 1.1 shows the ORB and the different interface categories described in the reference model.

1.3.2   ORB

The ORB is the central component of the OMA that glues all the other components together. The *Common Object Request Broker: Architecture and Specification* [2] defines the programming interfaces to the ORB. Note that all OMG specifications define objects just in terms of IDL interfaces and their semantics and not in terms of their particular implementation. This allows CORBA vendors considerable flexibility in the design of their particular implementation, which is important because different environments often pose specific constraints and requirements on the ORB. For example, ORB implementations for mainframe computers will be quite different from the ones that reside in embedded systems. But despite differences in the actual implementation, the CORBA specifications ensure that all ORBs can communicate with each other.

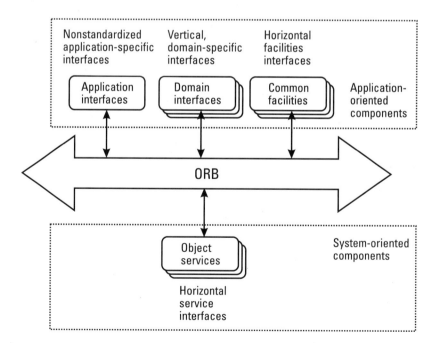

Figure 1.1 CORBA/OMA reference model interface categories.

An ORB is the basic mechanism by which objects transparently make requests to—and receive responses from—each other, either on the same machine or across the network. Clients do not need to be aware of the communications mechanisms, object activation, object implementation, and object location. The ORB thus forms the foundation for building applications constructed from distributed objects and for interoperability between applications in both homogeneous and heterogeneous environments.

The OMG IDL provides a standard way to define the interfaces to CORBA objects. IDL is a strongly typed language that is programming language–independent. Language mappings allow IDL interfaces to be implemented in the developer's programming language of choice in a style that is natural to that language (see Section 1.4.3).

1.3.3   Object Services

Object services are domain-independent (i.e., horizontally-oriented building blocks that are either fundamental for developing distributed CORBA applications or that provide a universal basis for application interoperability). In other words, object services provide system-oriented functionality to CORBA applications by allowing application developers to call object service functions instead of writing and calling their own private object service functions.

The OMG *CORBAservices: Common Object Services Specifications* [5] defines a number of different CORBA services that are briefly described in the following:

- The *naming service* provides the ability to bind a name to an object relative to a naming context[1] (like a telephone directory). A naming context is an object that contains a set of name bindings in which each name is unique. The underlying model is very general and flexible, since the component attribute values are not assigned or interpreted by the naming service.

- The *event service* provides asynchronous interactions between anonymous objects. It supports asynchronous events, event "fan-in," notification "fan-out," and reliable event delivery. The design is scaleable and suitable for distributed environments. No centralized

---

1. A *naming context* is fundamentally different from a *security context*. A *naming context* binds a name to an object to make it possible to locate it on the network. A *security context* provides both client and target side with the same security information.

server or global service is required. Both pull and push event delivery models are supported. Suppliers can generate events without knowing the identities of consumers, and consumers can receive events without knowing the identities of the suppliers. Event suppliers, consumers, and channels are objects.

- The *notification service* enhances the event service by supplying not only the event filtering features, but also structured event types and varying degrees of control over the quality of service that an event channel provides.

- The *life cycle service* deals with life, death, and relocation of objects. It defines operations to copy, move, and remove graphs of related objects (see also relationship service).

- The *persistent object service (POS)* allows objects to "persist" beyond the application that creates the object or the clients that use it. POS allows the state of an object to be saved in a persistent store and restored when it is needed. The object is responsible for managing its state but can use or delegate to the POS for the actual work. There can be a variety of different clients and implementations of the POS, and they can work together.

- The *object transaction service (OTS)* defines interfaces that allow multiple, distributed objects to cooperate in order to provide atomicity. These interfaces enable the objects to either commit all changes together or to rollback all changes together, even in the presence of (noncatastrophic) failure. It supports multiple transaction models, including the flat (mandatory) and nested (optional) models. It also supports interoperability between different programming models and different systems, including the ability to have one transaction service interoperate with a cooperating transaction service using different ORBs.

  The OTS supports both implicit (system-managed transaction) and explicit (application-managed) propagation. With implicit propagation, transactional behavior is not specified in the operation's signature. With explicit propagation, applications define their own mechanisms for sharing a common transaction. The OTS can be implemented in a TP monitor environment to support the ability to execute multiple transactions concurrently and to execute clients, servers, and transaction services in separate processes.

- The *concurrency control service* enables multiple clients to coordinate their access to shared resources. Coordinating access to a resource means that when multiple, concurrent clients access a single resource, any conflicting actions by the clients are reconciled so that the resource remains in a consistent state.

    Concurrent use of a resource is regulated with locks. Each lock is associated with a single resource and a single client. There are several lock modes that correspond to different categories of access. This variety of lock modes provides flexible conflict resolution.

- The *relationship service* allows components and objects that know nothing of each other to be related. This can be done without changing existing objects or requiring that they add new interfaces. In other words, dynamic relationships between immutable objects can be created. The service keeps track of the relationships between objects; the related objects are not even aware that they are part of a relationship. The service defines two new kinds of objects: relationships and roles. Roles are the objects in a CORBA system.

- The *externalization service* defines protocols and conventions for externalizing and internalizing objects. Externalizing an object is to record the object state in a stream of data, which can then be internalized into a new object in the same or a different process. A stream is a data holding area with an associated cursor. A cursor is a pointer that moves forward and backward as you write and read data to and from a stream. The data holding area can be in memory, on a disk file, or across a network.

- The *licensing service* provides a mechanism for producers to control the use of their intellectual properties. Producers can implement the licensing service according to their own needs and the needs of their customers. The current trend is toward component licensing, in which components will have to be written to automatically register with license managers. The service lets you meter the use of your components in a flexible manner and charge accordingly.

- The *query service* allows users and objects to invoke queries on collections of other objects. The queries are declarative statements with predicates, and they include the ability to specify values of attributes. Several query languages can be used (e.g., SQL).

    The query service provides an architecture for a nested and federated service that can coordinate multiple, nested query evaluators.

- The *property service* provides the ability to dynamically associate named values with objects outside the static IDL-type system (i.e., properties are essentially typed, named values that can be dynamically associated with an object outside the IDL type system). For example, it is possible to add an archive property to an existing document at run-time and mark the document as ready to be archived. The archive information is associated with the object, but it is not part of the object's type.

- The *time service* enables users to obtain current time together with an error estimate associated with it. It ascertains the order in which events occurred and computes the interval between two events. It consists of two services that manage universal time objects (UTO), time interval objects (TIO), and timer event handler objects. Maintaining a single notion of time is important for ordering events that occur in distributed object systems.

- The *collections service* allows the user to manipulate objects in a group. Collections are groups of objects that, as a group, support some operations and exhibit specific behaviors that are related to the nature of the collection rather than to the type of objects they contain. Examples of collections are sets, queues, stacks, lists, and binary trees.

- The *trader service* allows users to discover objects based on the services they provide (like the yellow pages). Exporters advertise their services with the trader; importers use the trader to discover services that match their needs. The trading service in a single trading domain may be distributed over a number of trader objects. Traders in different domains may be federated.

- The *security service* comprises the following services: identification and authentication, authorization and access control, security auditing, security of communication, nonrepudiation, and administration. After reading this book, readers will have in-depth knowledge about the CORBA security service.

Note that only some usable implementations of these services are commercially available at the time of this writing, (for example, the naming service, event service, trading service, object transaction service, relationship service, and security service) but it is anticipated that the whole range of services will be available in the near future.

1.3.4 Common Facilities and Domains

Common facilities are a collection of services that many applications may
share but which are not as fundamental as the object services. Common
facilities fill the conceptual space between the enabling technology (defined
by the CORBA ORB and object services specification) and the application-
specific unstandardized services that the OMA labels as application objects.
For instance, a system management or electronic mail facility could be classified as a common facility. Common facilities are divided into two major
categories. The first category contains *horizontal common facilities*, which are
used by most systems. There are currently four major domains for those
facilities: user interface, information management, systems management, and
task management. The second category describes *vertical market facilities*,
which support domain-specific tasks associated with vertical market seg-
ments. Information about the architecture of common facilities is in *COR-
BAfacilities: Common Facilities Architecture* [3].

There are no clear boundaries between CORBA facilities and CORBA
services. The CORBA services were defined from the bottom up, based on
the perceived need for enabling interfaces and capabilities. The CORBA
facilities, on the other hand, are typically derived from top-down needs.
CORBA facilities are concerned with application interoperability and not
with infrastructure and portability issues, which are mainly the responsibility
of the lower-level CORBA services and ORB.

Domain interfaces [1] fill roles similar to object services and common
facilities, but are oriented toward specific application domains, such as
finance, health care, manufacturing, telecommunications, electronic commerce, and transportation. In Figure 1.1, multiple boxes are shown for
domain interfaces to indicate the existence of many separate application
domains.

1.3.5 Application Interfaces

Application objects are developed specifically for a given application. Application objects can be built using other, more basic objects, some of them specific to a particular application object and some taken from the common
facilities. Application objects are not standardized by the OMG.

For example, you could build a word-processing application object
using application-specific objects to handle the creation and editing of a
document, and common facilities objects to handle the printing and storage
of a document.

1.3.6   Object Frameworks

Figure 1.2 illustrates the other part of the OMA reference model, the concept of object frameworks. These are domain-specific groups of objects that interact to provide a customizable solution within that application domain (e.g., telecommunications, medical, finance, manufacturing). In Figure 1.2, each circle represents a component that uses the ORB to communicate with other components. The interfaces supported by each component are indicated.

Within an object framework like the one shown in Figure 1.2, each component communicates with others on a peer-to-peer basis. That is, each component is both a client of other services and a server for the services it provides. In CORBA, the terms *client* and *server* are merely roles that are filled on a per-request basis. Very often, a client for one request is the server for another. The circles represent components, some with only one interface category and others with multiple categories.

1.4   CORBA

One of the first specifications to be adopted by the OMG was the *Common Object Request Broker: Architecture and Specification* [2]. It details the interfaces and characteristics of the ORB component of the OMA. CORBA's main features and functional components are discussed in this section:

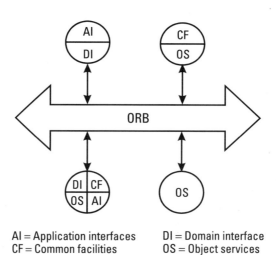

AI = Application interfaces   DI = Domain interface
CF = Common facilities   OS = Object services

Figure 1.2 OMA reference model interface use.

- ORB core;
- OMG IDL;
- Language mappings;
- Static invocation through stubs and skeletons;
- Interface and implementation repositories;
- Dynamic invocation and dispatch;
- Object adapters;
- Inter-ORB-protocols.

Most of these are illustrated in Figure 1.3, which also shows how the components relate to one another.

1.4.1 ORB Core

As mentioned above, the ORB component delivers requests to target objects and returns responses to the clients making the requests. The ORB's key feature is the transparency of client/object communication. The ORB hides the following:

- *Object location* (i.e., the client does not know where the target object resides).

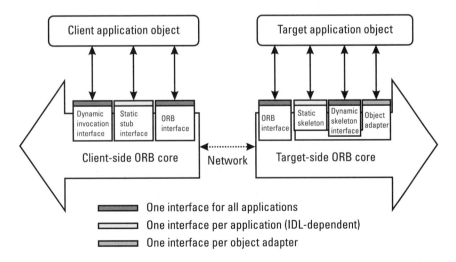

Figure 1.3 CORBA.

- *Object implementation* (i.e., the client does not know how the target object is implemented, what programming or scripting language(s) it was written in, or on what operating system and hardware it executes).

- *Object execution state* (i.e., the client does not need to know if the target object is currently activated or if it is in an executing process and ready to accept requests). The ORB transparently starts the object if necessary before delivering the request to it.

- *Object communication mechanism* (i.e., the client does not need to know what communication mechanisms the ORB uses).

These ORB features allow application programmers to focus more on their own application programming and less on low-level distributed system programming issues.

1.4.2 Object References

To make a request, the client specifies the target object by using an object reference that is automatically created when a CORBA object is created. An object reference always refers to the same object instance for which it was created, and the client cannot modify it. In other words, object references are both immutable and opaque. Object references are strongly typed and can have standardized interoperable formats. These references are called interoperable object references (IORs). Alternatively, they can have proprietary formats.

Conceptually, an IOR contains three parts of information:

- A standardized *repository ID*, a string that identifies the most derived type of the IOR at the time it was created. This makes it possible to locate a detailed description of the corresponding interfaces in the interface repository (see Section 1.4.6).

- Standardized *endpoint information* that is used by the ORB to establish the connection to the server identified by the IOR. It contains protocol information and physical addressing information.

- The ORB-proprietary *object key* that is used by the ORB to locate the object adapter, and by the object adapter to locate the servant that is to be invoked.

IORs that contain multiple endpoint information and object key fields are called multicomponent profiles, which allow IORs to support more than one protocol and transport mechanism. ORBs can use such multicomponent profiles to dynamically choose the protocol and transport depending on what the client and server support.

Object references can be obtained in three different ways:

- *At object creation:* A creation request returns an object reference for the newly created object to the client.[2]

- *Through a directory service:* A client can invoke a lookup service (e.g., naming service and trading service in *CORBAservices: Common Object Service Specifications*) in order to obtain object references for existing objects.

- *By converting references to strings and back:* An object reference can be converted into a string and stored into a file or a database. Even after being *stringified* and *destringified,* it can be used to make requests on the object as long as the object still exists.

1.4.3   OMG IDL

Before a client can make requests to an object, it must know the types of operations supported by the object. An object's interface specifies the operations and types that the object supports and thus defines the requests that can be made on the object. Interfaces for objects are defined in the OMG IDL. IDL interfaces are similar in syntax to classes in C++ and interfaces in Java.

The following exemplary IDL interface for a bank account illustrates the similarity between IDL and C++:

```
interface Account {
    void deposit( in unsigned long amount );
    void withdraw( in unsigned long amount );
    long balance();
};
```

Code Example 1: Example IDL interface.

---

2. Note that CORBA has no special client operations for object creation—generating objects is done by invoking creation requests, which are just ordinary operation invocations on other objects called factory objects.

  The fact that CORBA has no special object creation function or built-in directory service is indicative of a key theme of CORBA: Keep the ORB as simple as possible, and push as much functionality as possible to other OMA components such as object services and common facilities.

An important feature of OMG IDL is its language independence. Since OMG IDL is a declarative language, not a programming language, it forces interfaces to be defined separately from object implementations. This allows objects to be constructed using different programming languages and yet still communicate with one another.

The main features of the IDL type system are summarized below. For more information, see references [2, 6].

- Built-in types (e.g., `long`, `long long`, `short`, `float`, `double`, `long double`, `char`, `wchar`, `boolean`, `octet`, `enum`, `any`);
- Constructed types (e.g., `struct`, `union`);
- Template types (e.g., `string`/`wstring`, `sequence`, `fixed`);
- Object reference types;
- Interface inheritance.

1.4.4 Language Mappings

As mentioned above, OMG IDL is just a declarative language, not a full-fledged programming language. As such, it does not provide features like control constructs, nor can it directly be used to implement distributed applications. Instead, language mappings determine how IDL is mapped to the facilities of a given programming language. Important aspects of any language mapping are mappings for interfaces (and other pseudo-objects), types, and objects, to the corresponding constructs of the programming language that the IDL is mapped into (e.g., in C++: classes/functions, types, and programming language objects, respectively). As of this writing, CORBA standardizes language mappings for C, C++, COBOL, Java, Smalltalk, and Ada 95. Other language mappings also exist but have not yet been standardized by the OMG.

In practice, an IDL compiler automatically does the language mapping. It produces client-side code *stubs* and server-side code *skeletons* that form the basis for the actual implementation of the objects in the respective programming language. The details of these files, such as the names and number of generated source files, vary from ORB to ORB.

IDL language mappings are where the abstractions and concepts specified in CORBA meet the "real world" of implementation. Thus, their importance regarding CORBA applications cannot be overstated. A poor or incomplete mapping results in programmers being unable to effectively use CORBA technology in their language. Therefore, language mapping specifications are always undergoing periodic improvement in order to incorporate evolution of programming languages (e.g., Java), and add features that fulfill new requirements discovered by writing new applications.

1.4.5 Static Invocation: Stubs and Skeletons

In addition to generating programming language types, IDL language compilers and translators also generate client-side stubs and server-side skeletons. A stub is a mechanism that effectively creates and issues requests on behalf of a client, while a skeleton is a mechanism that delivers requests to the CORBA object implementation. A stub is essentially a proxy for the actual target object. Dispatching through stubs and skeletons is often called *static invocation*. IDL stubs and skeletons are built directly into the client application and the object implementation. Therefore, they both have to have complete a priori knowledge of the IDL interfaces of the objects being invoked.

A request sent by the client is first converted from the representation in the programming language to one that is suitable for transmission. Once the request arrives at the target object, the skeleton converts it to a (possibly different) representation—which depends on the underlying hardware and software platform—and dispatches it to the object. The response is sent back the way it came. Figure 1.3 shows the positions of the stub and skeleton in relation to the client application, the ORB, and the object implementation.

1.4.6 Interface and Implementation Repositories

The interface repository (IR) provides persistent objects that represent the IDL information in a form that is available for lookup at run-time. Using the information in the IR, it is possible for a program to be able to determine what operations are valid on an object and make an invocation on it, even if the interface was not known at compile-time. Using the IR interface of the IR object, applications can traverse an entire hierarchy of IDL information.

For example, an application can start at the root of the IR and iterate over all the module definitions there to search for the desired object. When the desired object is found, it can open the IDL file and iterate in a similar manner over all the definitions to retrieve information on interfaces and types. This hierarchical approach can be used to examine all the information stored within an IR.

Since the IR allows applications to programmatically discover type information at run-time, its real utility lies in its support of CORBA dynamic invocation (see Section 1.4.7). It can also be used as a source for generating static support code for applications (as described in Section 1.4.5).

The implementation repository contains information that allows the ORB to locate and activate implementations of objects. Ordinarily, installation of implementations and control of policies related to the activation and execution of object implementations is done through operations on the implementation repository. The implementation repository is also a common place to store additional information associated with implementations of ORB objects (e.g., debugging information). Figure 1.4 summarizes the use of the interface repository and implementation repository.

1.4.7 Dynamic Invocation and Dispatch

CORBA supports two interfaces for dynamic invocation: the dynamic invocation interface (DII), which supports dynamic client request invocation, and the dynamic skeleton interface (DSI), which provides dynamic dispatch to objects. The DII and DSI can be viewed as a generic stub and generic skeleton, respectively. Each is an interface provided directly by the ORB, and neither is dependent on the particular IDL interfaces of the objects being invoked.

The DII supports three types of requests:[3]

- *Synchronous invocation:* The client invokes the request and then blocks and waits for the response [similar to a remote procedure call

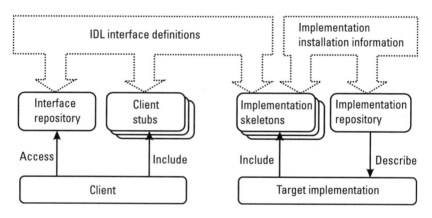

Figure 1.4 Interface and implementation repositories.

---

3. Currently, CORBA applications must use the DII for deferred synchronous invocation and one-way invocation. However, this restriction will soon be removed; an asynchronous messaging service is being developed.

(RPC)]. This is the most common invocation mode used for CORBA applications because it is also supported by static stubs.

- *Deferred synchronous invocation:* The client invokes the request, continues processing while the request is dispatched, and later collects the response. This is useful if long-running services are invoked.

- *One-way invocation:* The client invokes the request and then continues processing; there is no response.

While the DII offers more flexibility than static stubs, programmers should be aware of its potential hidden costs. DII is often slow because the interface repository has to be queried for each object invocation, and this often requires a transparent request to a remote location. Static invocations do not suffer from the overhead of accessing the IR since they rely on type information already compiled into the application.

Analogous to the DII is the server-side DSI,[4] which allows servers to be written without having skeletons for the objects compiled statically into the program. In real-world applications, the DSI is rarely used.

1.4.8   Object Adapters

Object adapters serve as the glue between CORBA object implementations and the ORB itself. In other words, an object adapter is an interposed entity that allows a caller to invoke requests on an object even though the caller does not know that object's true interface. Figure 1.5 illustrates the role of an object adapter.

CORBA object adapters are responsible for creating object references. They also ensure that each target object is incarnated by a servant, and they pass requests from the server-side ORB to the target servant. The functionality provided by the ORB through an object adapter often also includes interpretation of object references, method invocation, security of interactions, object and implementation activation and deactivation, mapping object references to implementations, and registration of implementations.

CORBA without object adapters would mean that object implementations would need to connect themselves directly to the ORB to receive requests. Instead of object adapters, a very complex ORB interface would be

---

4. Unlike most of the other CORBA subcomponents, which were part of the initial CORBA specification, the DSI was only introduced at CORBA 2.0.

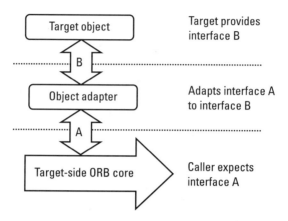

Figure 1.5 Role of an object adapter.

required, which would be difficult to standardize. Without object adapters, the ability of CORBA to flexibly support diverse object and ORB implementations would be severely compromised.

CORBA allows for multiple object adapters per ORB, addressing the wide range of object granularities, lifetimes, policies, and implementation styles. Until Version 2.1, CORBA only specified one object adapter, the basic object adapter (BOA). CORBA required that a BOA be available in every ORB, and it specified the functions that had to be provided in a BOA implementation. When it was first specified, it was hoped that the BOA would suffice for the majority of object implementations. As a result of the goal to make the BOA support multiple languages, the specification had to be made quite vague in some areas. This, in turn, resulted in nontrivial portability problems between BOA implementations because ORB vendors had to fill in the missing pieces with proprietary solutions.

Since Version 2.2, CORBA specifies the portable object adapter (POA) that replaced the flawed BOA. The POA supports the whole range of interactions between CORBA objects and programming languages while at the same time maintaining application portability. Therefore, the BOA specification has been removed from CORBA, and the POA is the new standard object adapter.

1.4.9   Inter-ORB Protocols

CORBA 1.1 was only concerned with creating portable object applications. CORBA 2.0 introduced a general ORB interoperability architecture that answered the demand for direct ORB-to-ORB interoperability and for bridge-based interoperability. Direct interoperability is possible when two ORBs reside in the same domain—in other words, both ORBs understand the same object references, use the same OMG IDL type system, communicate via the same low-level protocol, and perhaps share the same security information. Bridge-based interoperability means that ORBs from separate domains must communicate. The role of such bridges is to map ORB or domain-specific information from one ORB domain to another.

The ORB interoperability architecture is based on the abstract General Inter-ORB Protocol (GIOP), which specifies transfer syntax and a standard set of message formats for ORB interoperation over any connection-oriented transport. The Internet Inter-ORB Protocol (IIOP) specifies how GIOP is built over TCP/IP transports. Every ORB that calls itself CORBA-compliant must either implement IIOP natively or provide a half-bridge to it. IIOP is by far the most commonly used CORBA protocol today. The ORB interoperability architecture also provides for other environment-specific inter-ORB protocols (ESIOP) that allow ORBs to be built for special situations in which certain distributed computing infrastructure is already in place.[5] For example, the first ESIOP adopted was the DCE Common Inter-ORB Protocol (DCE-CIOP) that can be used by ORBs in environments where DCE is already installed.

In addition to standard interoperability protocols, standard object reference formats are necessary for ORB interoperability. CORBA specifies a standard object reference format called the IOR. An IOR stores information needed by the ORBs to locate other objects and communicate with them over one or more protocols (see Section 1.4.2). For example, an IOR containing IIOP information stores host name, TCP/IP port number, and other required information.

1.5 How Does It All Work Together?

This section tries to put all CORBA components into context. A CORBA product generally consists of the following components:

---

5. Note that both the IIOP and DCE/ESIOP have built-in mechanisms for implicitly transmitting context data that is associated with the transaction or security services.

- The ORB library that gets linked into the application code and acts as a proxy for all remote method invocations. This library contains the implementation for the CORBA protocols.

- Some CORBA implementations use some kind of activation daemon that listens for requests to the host on which it resides, and restarts the called target ORB and target application object. The advantage is that not all target ORB and application objects on the host need to run on a permanent basis.

- The IDL compiler that provides the language mapping from the standardized object interfaces into matching client stubs and target skeletons. These stubs and skeletons form the basis of the application code (i.e., the application programmer implements the actual object functionality within the stubs and skeletons).

- Add-on CORBA services such as a naming service, a transaction service, or a security service can be installed to enrich CORBA's middleware functionality. These products are often provided by highly specialized third-party vendors.

Figure 1.6 shows the main components of CORBA and how they are interrelated. The following sections offer a brief explanation of the different APIs used during the CORBA invocation process.

1.5.1   The Client System at Run-Time

Initially, a client application invokes operations on some server application object. The client can either use the stub-style invocation API (client stub) or the dynamic invocation API (dynamic invocation interface) to invoke the operation either statically or dynamically, respectively. Logically, the ORB is a single component, but it has some functions specific to the client side and other functions specific to the server side. The client-side ORB handles the invocation request from the client and selects the related servers and methods. It validates arguments against the interfaces and sends the request to the server side ORB or activation component. The application can invoke methods synchronously or asynchronously. The ORB is linked into the client application itself. The interface repository stores modules of interface information (object references), including descriptions of the operations that are valid for a given object and the arguments that are valid for an operation. Context objects contain information about the client environment, or a request that is not passed as formal arguments of an operation. CORBA

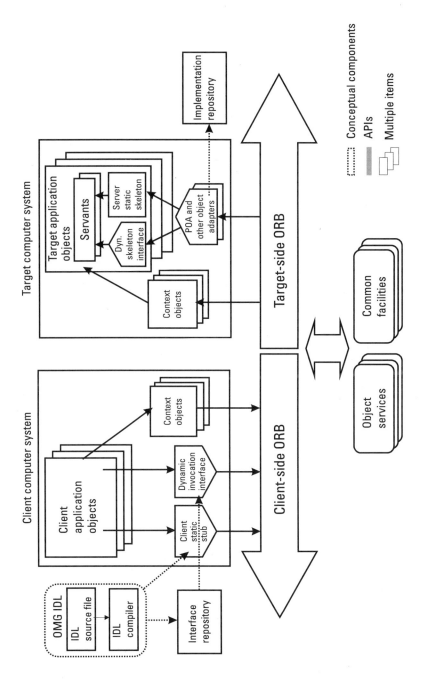

Figure 1.6 An architectural overview of CORBA.

transmits context object information (a list of properties and their values) from the client to the server, and potentially to other servers.

1.5.2   The Server System at Run-Time

The server-side ORB, which is linked into the server application, receives the method dispatch request, unmarshals the arguments, sets up the context state as needed (i.e., context objects), invokes the method dispatcher in the server skeleton, and completes the invocation. The ORB API enables application developers to access all ORB functions that do not depend on a specific object adapter, such as for manipulating object references. The object adapter API enables method developers to access CORBA functions, such as registering implementations, authenticating requests, and handling activation policies. Object adapters perform general ORB-related tasks, for example, activating objects and implementations and registering server instances. The POA is the most commonly used object adapter. The server skeleton takes these general tasks and ties them to particular implementations and methods in the server. A server skeleton is a language-specific mapping of OMG IDL operation definitions into methods. For dynamic invocations, the DSI API can be used instead of the server skeleton. The server application communicates with the server-side ORB and includes one or more implementations of an object. The method implementations (called *servants*) are the parts of a server application that satisfy a client's request for an operation on a specific object. The implementation repository is a storage place for implementation definitions, such as information about which implementations are installed on a given system.

1.5.3   Other Components

Stubs and skeletons are generated by the IDL compiler. Object references are created by the object adapter when the object is generated. Object services provide a set of run-time system level services (see Section 1.3.3). Object facilities (see Section 1.3.4) are higher-level services than the object services.

1.6   Creating and Running an Application Example

This section will introduce a simple example application and briefly illustrate how the IDL interface is specified, how the stubs and skeletons are compiled, how the client and server objects are implemented from the stubs and skeletons, and, finally, how the example is executed. Detailed information about the code example that is described can be found in the documentation that comes with the MICO CORBA distribution [7].

1.6.1 Writing a CORBA Application

The first step for the programmer is developing the IDL interfaces for all application objects. As a general rule, an IDL interface should provide as little access as possible to internal functions and variables, without inhibiting the application's functionality.

Consider an example application in which a bank would maintain its customers' accounts. An object that implements such a bank account should offer the following three operations: deposit a certain amount of money, withdraw a certain amount of money, and provide a balance that returns the current account balance. The state of an account object consists of the current balance. The following IDL file `account.idl` with two interfaces captures that functionality:

```
interface Account {
    void deposit( in unsigned long amount );
    void withdraw( in unsigned long amount );
    long balance();
};

interface Bank {
    Account create ();
};
```

Code Example 2: `account.idl`

The next step is to run this interface declaration through an IDL compiler that will generate code in the target programming language. The MICO IDL compiler is called `idl` and is used like this to produce C++ code stubs and skeletons that run with the POA:

```
idl —poa —no-boa account.idl
```

The IDL compiler will generate two files: `account.h` and `account.cc` that contain the class declarations for the account implementation base class (`POA_account`) and the client stub (`account_stub`). The `POA_` prefix in the implementation base class name indicates that the POA is used to connect the target object to the target side ORB. account.h contains class declarations for the base class of the account object implementation and the stub class a client will use to invoke methods on remote account objects. account.cc contains implementations of those classes and some supporting code.

The application programmer now needs to subclass POA_account (implementing the purely virtual methods) and write a program that uses the bank account object.

The following first code fragment shows a simple implementation of the methods provided by the target servant object. The *.h files that describe the ORB and all other CORBA declarations are automatically included through account.h. This part of the target object is called a *servant*, which is not the same as a server. A servant contains the actual interface implementation, whereas the server contains the code to bootstrap the system.

```
#include <fstream.h>
#include "account.h"

/*
 * Implementation of the Account
 */

class Account_impl : virtual public POA_Account
{
public:
  Account_impl ();

  void deposit (CORBA::ULong);
  void withdraw (CORBA::ULong);
  CORBA::Long balance ();

private:
  CORBA::Long bal;
};

Account_impl::Account_impl ()
{
  bal = 0;
```

```
}

void
Account_impl::deposit (CORBA::ULong amount)
{
  bal += amount;
}


void
Account_impl::withdraw (CORBA::ULong amount)
{
  bal -= amount;
}

CORBA::Long
Account_impl::balance ()
{
  return bal;
}

/*
 * Implementation of the Bank
 */

class Bank_impl : virtual public POA_Bank
{
public:
  Account_ptr create ();
};

Account_ptr
Bank_impl::create ()
{
  /*
   * Create a new account (which is never deleted)
   */

  Account_impl * ai = new Account_impl;
```

```
/*
 * Obtain a reference using _this.
 * This implicitly activates the
 * account servant (the RootPOA, which is
 * the object's _default_POA,
 * has the IMPLICIT_ACTIVATION policy)
 */

Account_ptr aref = ai->_this ();
assert (!CORBA::is_nil (aref));

/*
 * Return the reference
 */

return aref;
}
```

Code Example 3: server.cc (part 1)

From an application developer's perspective, this CORBA-enabled application code does not differ much from a normal program—apart from the inheritance and a few simple naming conventions. This is why using CORBA is rather easy for application programmers.

Next is the main part of the server application (this component is called the server), which starts the ORB, creates and starts a Bank object, and writes the object reference to the object into a file Bank.ref. There are many ways in which the object reference can be transferred from the target side to the client side, in particular via a naming service or trading service. However, this simple example just uses a file to transmit the reference so that no additional CORBA components are required.

Please note that the example uses the POA as a connection mechanism between the target ORB and target implementation.

```
int main (int argc, char *argv[])
{
  /*
   * Initialize the ORB
```

```
*/

CORBA::ORB_var orb = CORBA::ORB_init (argc, argv);

/*
 * Obtain a reference to the RootPOA and its Manager
 */

CORBA::Object_var poaobj =
   orb->resolve_initial_references ("RootPOA");

PortableServer::POA_var poa =
   PortableServer::POA::_narrow (poaobj);

PortableServer::POAManager_var mgr =
   poa->the_POAManager();

/*
 * Create a Bank
 */

Bank_impl * micocash = new Bank_impl;
/*
 * Activate the Bank
 */

PortableServer::ObjectId_var oid =
   poa->activate_object (micocash);

/*
 * Write reference to file
 */

ofstream of ("Bank.ref");
CORBA::Object_var ref =
   poa->id_to_reference (oid.in());

CORBA::String_var str =
   orb->object_to_string (ref.in());
```

```
of < str.in() < endl;
of.close ();

/*
 * Activate the POA and start serving requests
 */

printf ("Running.\n");

mgr->activate ();
orb->run();

/*
 * Shutdown (never reached)
 */

poa->destroy (TRUE, TRUE);
delete micocash;

return 0;
}
```

Code Example 3: server.cc (part 2)

Now the target object is up and running, and the object reference that points to that object is saved in a file.

The following client application reads the object reference from the file Bank.ref, binds to the server, opens a new account, deposits 700, withdraws 250, and displays the result. Again, the *.h files that describe the ORB and all other CORBA declarations automatically included through account.h.

```
#include "account.h"

#ifdef HAVE_UNISTD_H
#include h
#endif
```

```
int main (int argc, char *argv[])
{
  CORBA::ORB_var orb = CORBA::ORB_init (argc, argv);

  /*
   * IOR is in Bank.ref in the local directory
   */

  char pwd[256], uri[300];
  sprintf (uri, "file://%s/Bank.ref",
      getcwd(pwd, 256));

  /*
   * Connect to the Bank
   */

  CORBA::Object_var obj = orb->string_to_object (uri);
  Bank_var bank = Bank::_narrow (obj);

  if (CORBA::is_nil (bank)) {
    printf ("oops: could not locate Bank\n");
    exit (1);
  }

  /*
   * Open an account
   */

  Account_var account = bank->create ();

  if (CORBA::is_nil (account)) {
    printf ("oops: account is nil\n");
    exit (1);
  }

  /*
   * Deposit and withdraw some money
   */
```

```
account->deposit (700);
account->withdraw (450);

printf ("Balance is %ld.\n", account-balance ());
return 0;
}
```

Code Example 4: `client.cc`

More information about this code example can be found in the documentation that comes with the MICO CORBA distribution [7]. Throughout this book, the functionality of this basic application will be extended step-by-step with functional components of the CORBA security service.

1.6.2 Running a CORBA Application

After the client and target code have been successfully built, the server application needs to be launched on the server machine and the object reference for the server needs to be transferred to the client machine.

The client application can then be launched. It will read the object reference for the server application from the file and use it to locate the server application and bind to it. Finally, the server object can be invoked by the client application just like a local object.

1.7 Summary

CORBA, an industry specification that has been in development since 1990 by the OMG, allows software objects to talk to each other across the network. It has many advantages. From a business perspective, CORBA can help with legacy integration, which leverages the use of both old and new systems. From a technical point of view, CORBA can provide transparency, platform independence, portability, software reuse, integration, interoperability, flexibility, and scalability.

CORBA is one part of the OMA, which is an umbrella architecture for all OMG specifications. The main component of the OMA is the ORB that glues all the other components on the network together. To achieve this, it uses system-level object services such as the naming service or security service. This book is about the CORBA security service. Common facilities and domains are similar to object services but more directed at specific application needs.

CORBA specifies object interfaces in its own IDL to enable platform and programming language independence. These IDL interface specifications are compiled into stubs and skeletons in the underlying programming language—the application programmer then implements the actual object functionality using these stubs and skeletons. On the server side, object adapters such as the POA serve as the glue between CORBA object implementations and the ORB itself.

The ORB uses object references as pointers to invoke target objects. Invocation can be done statically or dynamically. Dynamic invocation often uses repositories that hold interface and implementation information needed to locate and identify the invoked object. Remote ORBs communicate through a suite of network-independent CORBA-specific inter-ORB protocols (e.g., the IIOP).

This chapter also contains an example run-through of a CORBA invocation to illustrate how all these different components work together. A code example of a simple banking application gives a brief introduction to the application development process using CORBA.

1.8 Further Reading: Books on CORBA

This book is about CORBA security and not about CORBA. Readers should therefore draw information from other books if they want to know more about the inner workings of CORBA and other OMA components.

As of this writing, by far the most comprehensive book on CORBA programming is *Advanced CORBA Programming with C++* [6]. It provides in-depth technical advice and is a good desk reference on the topic. If readers prefer a light and entertaining read, then try *Instant CORBA* [8], which is a nontechnical description of CORBA and its components. Other books worth reading are *CORBA 3: Fundamentals and Programming* [9] and *Enterprise Application Integration with CORBA* [10].

References

[1] OMG, *CORBA Domain Technology: Manufacturing, Med, Finance, Telecoms,* 1998–2001.

[2] OMG, *Common Object Request Broker: Architecture and Specification,* Revision 2.5, September 2001.

[3]   OMG, *CORBAfacilities: Common Facilities Architecture,* http://www.omg.org.

[4]   OMG, *A Discussion of the Object Management Architecture,* Rev. 3.0, June 1997.

[5]   OMG, *CORBAservices: Common Object Services Specifications,* http://www.omg.org, 2001.

[6]   Henning, M., and S. Vinoski, *Advanced CORBA Programming with C++,* Reading, MA: Addison-Wesley, 1999.

[7]   MICO, *MICO is CORBA—An Open Source CORBA 2.3 Implementation,* Version 2.3.1, 2000.

[8]   Orfali, Robert, Dan Harkey, and Jeri Edwards, *Instant CORBA,* New York: Wiley, 1997.

[9]   Siegel, J., *CORBA 3: Fundamentals and Programming,* Second Edition, New York: Wiley/OMG, 2000.

[10]   Zahavi, R., *Enterprise Application Integration with CORBA,* New York: Wiley/OMG, 2000.

2

The Security Basics

2.1 What Is Security?

Most CORBA people have a hunch regarding what security means in their systems. This gut feeling is frequently based on urban myths that involve wily hackers who spend their days attacking computer systems, and some kind of magic cryptography that can be installed to spoil their malicious attempts. But this is far from the useful precise definition that is needed in order to understand what CORBA security is supposed to do. Hence, this chapter gives a concise characterization of key security concepts and terminology within the context of CORBA environments.

Security is about the protection of assets [1]. *Assets* can be tangible (e.g., network elements and hosts) or intangible (e.g., information or use of resources). In either case, the owner of each asset associates a value with it and, therefore, wants to restrict access to it.

CORBA security concerns information security—other types of security, such as physical security, cannot be achieved by CORBA security and are, therefore, considered outside this book's scope. The term *information security* is used to describe the task of preventing information assets from being compromised. Depending on the environment, this can include attacks on stored information, as well as information on the wire. Note that data is not the same as information—data is rather a representation of information. Further, information is the (subjective) interpretation of data. Data as such often has no value—it is the information conveyed through the data that has to be protected instead. Hansen [2] and other sources define data as "physical phenomena chosen by convention to represent certain aspects of our conceptual and real world. The meanings we assign to data are called information. Data is used to transmit and store information and to derive new information by manipulating the data according to formal rules."

In a similar sense, the ultimate goal of CORBA security is to protect information resources rather than objects, invocations, or messages. Applying security to these technical components is rather a means to an end, which is the protection of the information resources on the system.

Protective countermeasures can be distinguished in three categories [1]:

- *Prevention countermeasures:* Prevent assets from being damaged before the attack happens. A real-life example of prevention would be the locks on the doors of your house.

- *Detection countermeasures:* Allow you to detect when an asset gets compromised, giving you information on how the damage has been caused and who caused it. A burglar alarm in your house is an example of a physical detection measure.

- *Reaction countermeasures:* Allow you to recover the damaged assets or recover from damage to your assets after they have occurred. For example, you could call the police to find the burglar.

Each protective measure also has a cost associated with it, which is mainly based on the effort required to implement the security measures in the target environment. In the examples given above, these costs would be the expenses of buying and installing locks and burglar alarms and the effort spent by police.

2.2 Why Security?

Protection of assets is important because each asset has some value associated with it that can be lost when the asset is compromised. Tangible assets are often worth their reselling or buying price, which is generally relatively easy to determine. Only in some cases is the value of intangible assets as easy to determine, for example, for a service in which customers get charged on a per-bandwidth or per-timeslot basis, so that each "stolen" unit corresponds to a defined monetary value. For most other information assets, it is often hard to determine or quantify the exact value. This can be due to the fact that the value of some information has no exact financial equivalent or because the potential value is consequential. Examples of information assets that have a high consequential value are product designs or company strategies.

It is therefore more appropriate to quantify the *loss cost* as the value of assets. The loss cost can be defined as the negative impact when an asset is compromised: replacement cost, reputational damage, loss of competitive advantages, and loss of customers.

The purpose of security is to minimize the overall loss associated with potential attacks. By nature, there is no such thing as total security. Any system that provides valuable information resources to authorized users is vulnerable in one way or another because it reacts to user actions. Therefore, effective security enforcement is about finding an appropriate trade-off between the cost of implementing a security measure and the loss cost of a successful break-in. This process is often referred to as *risk analysis* (more on this topic can be found in Section 2.4.2).

In most commercial environments, the cost of implementing security measures needs to be lower than the total loss cost associated with all potential attacks (taking into account the chances of such attacks being successful). Otherwise, it would be more cost-effective to pay for the loss instead of the cost of implementing security measures. However, in some environments, in particular in the military, the definitions of cost and loss can be somewhat counterintuitive—it may be preferable to destroy some information asset instead of saving it if that prevents it from getting into the hands of the enemy. In this case, the loss is not defined by the loss of information but by the fact that it is valuable because it is unknown to other parties.

2.3 Security Properties Within CORBA

This section extends our definition of information security by distinguishing a number of different aspects of information security. We first give a generic classification of the main aspects involved and then describe a number of additional requirements for the CORBA security architecture. The discussion will show that there is no single universal terminology that fits all systems. Therefore, we try to identify a set of CORBA specific definitions.

Protection of information assets from unauthorized attempts to access information or from interference with its operation is often defined as having the following fundamental goals (as defined in the *Common Criteria* standard [3]):

- *Confidentiality:* The prevention of unauthorized disclosure of information;

- *Integrity:* The prevention of unauthorized modification of information;

- *Availability:* The prevention of unauthorized withholding of information or resources.

These goals are prioritized differently depending on the application and particular environment. In many commercial systems, availability is most important, whereas in some military environments, confidentiality may be more critical. One way of looking at these three main goals is that they all describe different aspects of access control to information resources. One can also argue that the list of fundamental goals is incomplete [1] and add accountability (e.g., if your applications provide e-commerce services where irrevocable evidence of actions has to be kept). *Dependability* is another important requirement related to the reliability of a (security) system in the face of attacks.

2.3.1 Confidentiality

Historically, security and confidentiality were closely related, because in the early days of computing, information security was mainly a concern to the military. As a result of the importance of confidentiality in the past, research in computer security has often concentrated on this topic. Even today, many people still feel that the main purpose of information security is to stop unauthorized users from reading (and understanding) sensitive information. Sometimes the terms *security* and *confidentiality* are even used synonomously, despite the fact that securing a system includes many other aspects. In this book, confidentiality is considered one part of security.

Confidentiality is enforced by restricting read operations, for example, by encrypting communications or by implementing access controls. Somewhat counterintuitively, research in information security has found that it is often also necessary to police write operations to enforce confidentiality. The Bell-LaPadula model [4] illustrates this phenomenon—confidentiality can be breached if privileged users who have access to confidential information are able to write it somewhere else where it is accessible to users who should not have access to it.[1]

For CORBA security, the following narrow definition of confidentiality is most appropriate: Confidentiality denotes the protection of requests

---

1. Note that this requires the intent of the privileged user to do so.

and replies from unauthorized reading, as well as the restriction of access to target operations that return confidential information, so that only authorized callers can invoke them.

2.3.2 Integrity

In a narrow sense, integrity deals with the prevention of unauthorized writing. In this interpretation, integrity is the dual of confidentiality. Oftentimes, similar security mechanisms achieve both goals.

However, it is not easy to give a concise definition of integrity. There are, in fact, several differing definitions. In the paper by Clark and Wilson [5], integrity means that no user of the system, even if authorized, may be permitted to modify data items in such a way that assets or accounting records of the company are lost or corrupted. The *Orange Book* [6] defines data integrity as external consistency (i.e., the state that exists when computerized data is the same as that in the source document and has not been exposed to accidental or malicious alteration or destruction). In communications security, integrity refers to the detection and correction of modification, insertion, deletion, or replay of transmitted data, including both intentional manipulations and random transmission errors [1].

For CORBA security, the latter definition of communications integrity is most appropriate. In addition, access to target operations that can modify protected information needs to be included in the definition.

2.3.3 Availability

The ISO/OSI 7498-2 security architecture for communications security [7] defines availability as the property of being accessible and useable on demand by an authorized entity. From a security perspective, an availability compromise is the prevention of authorized access to resources or delaying of time-critical operations. It ensures that a malicious attacker cannot launch so-called *denial of service attacks* (i.e., prevent legitimate users from having reasonable access to their systems).

This definition makes sense for CORBA security, although the CORBA security services [8] are themselves not able to fully provide protection from denial of service attacks. This is because availability is very much a concern beyond the traditional boundaries of information security. It involves engineering techniques that come from areas like fault-tolerant or real-time computing.

Availability is often partly the responsibility of other OMA components, such as the archive/restore services, or of underlying network or operating systems services. If the underlying network is vulnerable to denial of service attacks, then the CORBA security services that use the network cannot preserve availability. However, CORBA security can protect target implementations from denial of service attacks by only forwarding authorized requests to the application layer.

2.3.4   Accountability

The three goals of information security described so far can be interpreted as three different aspects of the same goal: to prevent unwelcome events by controlling access to resources. However, things can go wrong in any real-world system, no matter how effective the access controls are. For example, authorized actions can lead to security breaches, or bugs in the security system can be exploited by attackers. Therefore, we add another aspect to our definition of security that holds users responsible for their actions. The *Orange Book* [6] describes accountability as follows: Audit information must be selectively kept and protected so that actions affecting security can be traced to the responsible party. To achieve this, the system needs to authenticate users through an identification and authentication process, and it needs to keep audit logs of security relevant events associated with each user.

The CORBA security services specify two functional components to achieve accountability: security audit, which logs security relevant events (like an unsigned purchase receipt), and nonrepudiation, which generates irrevocable evidence of user actions (like a signed pay-slip). Although the general definition of accountability fits CORBA security, it is not clear which information can and should be audited in CORBA-based systems. CORBA security specifies the logging of specific audit events both on a per-invocation granularity on the ORB layer and more fine-grained on the application layer.

2.3.5   Dependability

Security systems have to deal with situations in which proper performance is required in the face of adverse conditions. The term *reliability* is concerned with the reaction of an IT system to failures, whereas *safety* relates to the impact of systems failures on their environment. IFIP WG10.4 [9] describes dependability as a unifying concept—security, reliability, integrity, and availability are simply aspects of dependability. Dependability is defined as the property of a computer system, such that reliance can justifiably be placed on the service it delivers to its users [10].

Dependability is a desirable property for any security system, because users have to rely on the security system to enforce the required protection measures in the intended manner. However, addressing dependability in CORBA security is hard because several layers play together to enforce the security measures, and dependability heavily depends on the layers below CORBA. To achieve dependability, CORBA security ultimately relies on the security functionality provided on the lowest possible layer, which is considered to be mostly hidden from the CORBA layer.

On a more general note, dependability of a security system increases the trustworthiness of the system. The CORBA security services specification [8] defines trustworthiness as the ability of a system to protect resources from exposure to misuse through malicious or inadvertent means.

2.4 Security Management

Security in CORBA systems is a difficult technical problem, but it is not just a technical issue. In many cases, bad security management is the real cause for security problems; rather than technical weaknesses. To help you prevent that, this section will explore, in general terms, how information security is managed on an organizational level. It is important to approach security and security management within the context of the goals and constraints of the environment as a whole. Otherwise, the security "solution" may not solve the real problems.

Good security management has several objectives: the general definition of the enterprisewide security goals (policy); the identification, implementation, and documentation of appropriate countermeasures (risk analysis); and the evaluation of countermeasure effectiveness based on analysis and feedback (audit). Figure 2.1 shows the connections between the different key documents, processes, and parties related to security management, followed by a more detailed description.

2.4.1 Security Policy

Most literature defines the term security policy rather vaguely—at times, there are even conflicting notions. In practice, it makes sense to define security policies on several hierarchical layers of abstraction, more static and

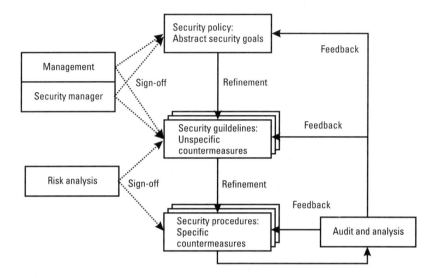

Figure 2.1   Security management overview.

abstract at the top, and with increasing level of detail and more regular modifications toward the lower layers.

From an organizational point of view, the information security policy specifies the abstract security goals within the context of the overall enterprise objectives, such as optimal long-term profit, adherence to data protection requirements, maintenance of good reputation. Its main objective is to provide management direction and support for information security [11].

On a technical level, the term *security policy* is often defined as a set of rules that state which actions are permitted and which actions are prohibited. A *domain* of a security policy is the set of entities (i.e., users, data objects, machines, administrators) that are governed by the policy [1]. Security rules are defined by an authorized entity, such as the security manager or administrator. The CORBA security services specification [8] defines the terms at an even more technical level: "A security policy is the data that defines what protection a system's security services must provide. There are many kinds of security policies, including an access control policy, audit policy, message protection policy, and nonrepudiation policy. Security policy domain is defined as a domain whose objects are all governed by the same security policy. There are several types of security policy domains, including access control policy domains and audit policy domains."

It is useful to define separate terms for security policies on different levels of abstraction. In this book, we will use the widespread approach of calling the abstract high-level document the "information security policy," the less abstract refinement of this policy the "security guidelines," and the resulting detailed mechanism-specific instructions "security procedures."

2.4.1.1 Information Security Policy

In essence, on a few pages, the information security policy document states what security is to be achieved. It defines the organization's fundamental goals regarding information security, where the responsibility for security lies, the organization's commitment to security, and the scope to which the security policy should apply. It should cover the security goals from a technical, social, legal, and management perspective, and ensure that the organization's security aims fit with the overall enterprise management strategy. The content should be specific enough to provide a useful framework for lower-level documents, and at the same time be abstract enough to remain relatively static over time. Modifications to this document should only be necessary in exceptional cases, such as major changes in the system environment.

Most enterprise security policies start with general statements like: "All departments should ensure that adequate information security management policies are implemented to protect their information assets." By signing off these declarations, the enterprise management has to commit itself to support the information security policy, both financially and through proactive management. The policy should then clarify critical issues such as asset ownership, responsibilities, minimum security requirements, personnel security, staff/user education, physical security, incident response/reporting/recovery, auditing, policy revision, and compliance with regulations. For example, a policy could state the requirement that access to information and resources should be restricted to authorized personnel.

There are a number of resources available that aid in the development of an enterprise security policy. The British standard BS 7799 [11] contains a set of "best practices" guidelines for information security and, more importantly, how organizations can demonstrate compliance to independent accredited auditors and receive certification. First published in early 1995, it was the first set of guidelines by a standards body that could reasonably be implemented by both small and large businesses. BS 7799 was updated in 1999 to include controls for e-commerce, mobile computing, teleworking, and outsourcing. The standard addresses 10 key areas of information security management: security policy, security organization, assets classification and control, personnel security, physical environmental security, computer and network management, system access control, systems development and maintenance, business continuity planning, and compliance. Other useful resources are the *German IT Baseline Protection Manual* [12] and the *ISO/IEC TR 13335 Guidelines for the Management of IT Security (GMITS)* [13].

2.4.1.2   Security Guidelines

Security guidelines identify the countermeasures that should be implemented to accomplish the security goals stated in the information security policy. If no appropriate countermeasure can be found for a policy statement, or if the countermeasure cannot be implemented in an acceptable (i.e., cost-effective) way, then the security guidelines should state this explicitly. And just like the information security policy, the security guidelines should be signed off by the security management, which thus takes responsibility for the choice of countermeasures and their accordance with the enterprise security goals. Changes to the guidelines can occur from time to time but only in harmony with the goals set out in the information security policy.

An appropriate security guideline for the aforementioned access restriction example could be along these lines: "The enterprise restricts access to information and resources through the use of passwords, with appropriate length and complexity to prevent brute-force attacks; passwords are not to be shared to ensure clear accountability; passwords are not to be written down, and they will be renewed on a regular basis."

2.4.1.3   Security Procedures

Security procedures state in detail how the identified countermeasures are to be implemented within particular environments. Because of their detailed nature, security procedures only apply to the exact configurations and versions of the system components for which they were written. This means that, for each countermeasure, there can be several customized procedures, for example, one for each application, for each ORB product in use on a particular platform, or one per set of security mechanisms used. Modifications to these documents can occur frequently as product updates become available or bugs are detected and patched. Adjustments can be carried out by system administrators and programmers whenever they change the system, but they should only become officially valid after the security manager has signed off on them to ensure compliance with the security guidelines and with the overall information security policy.

An exemplary security procedure for the password example above would be: "Passwords are chosen by the user but must contain at least six characters, with at least two alphanumeric letters (minimum one capital, one lowercase), and at least one numeric; passwords are renewed automatically every 28 days; the last 10 passwords are stored and cannot be chosen again; passwords must go through a one-way function before they are transmitted or stored; the stored values are made resilient to dictionary attacks by using a salting function (applies to: CORBA Banking v1.0, MICOSec level 2, v1.0, Compaq iPAQ PocketPC H3600, Handheld Linux V0.4)."

2.4.2 Risk Analysis

The previous section stated that the security policy derives the abstract information security goals from the overall business goals, but it did not explain how guidelines and procedures are generated from that. Knowing what security the enterprise wants to achieve is often not sufficient to be able to develop effective security guidelines and procedures, because in many cases there is no a priori knowledge of the potential threats to the security of a particular system.

Therefore, an intermediate step is introduced at which all potential risks to the security goals are captured and compared to the effectiveness of corresponding countermeasures. This process is referred to as *risk analysis*—the procedure used to estimate potential losses that may result from system vulnerabilities and to quantify the damage that may result if certain threats occur [13]. The ultimate goal of risk analysis is to help select cost-effective safeguards that will reduce the residual risk to an acceptable level.

The term *risk* can be defined as the index of the threats and corresponding vulnerabilities in a system. If both the threat and vulnerability are considered severe, then the risk is high. If the threat is severe but the vulnerability is minor, then the risk may be medium or low. Risk can also be interpreted as a measure for the probability of a loss to occur.

Some sources divide the risk analysis process into two parts, risk assessment and risk management. *Risk assessment* is defined as the assessment of threats to, impacts on, and vulnerabilities of information and information processing facilities and the likelihood of their occurrence. *Risk management* is the process of identifying, controlling, and minimizing or eliminating security risks that may affect information systems at an acceptable cost [11].

A full risk analysis proceeds in several steps. Initially, the functionality of the analyzed system needs to be documented in detail before the actual risk assessment can be carried out. This ensures that the analyst fully understands the inner workings of the system and prevents undocumented changes to the system that could alter the results of the analysis. In addition to the functionality of the system, the documentation should contain a description of the technical and organizational environment, to prevent deployment in inappropriate environments.

This is followed by the risk assessment step that, based on the system description, identifies threats and vulnerabilities and estimates the resulting risk. It is important to capture as many potential threats as possible and to consider all possible causes, such as different attack scenarios and accidental improper system use. The list of identified threats can be divided into several categories (dependent on the application), such as client-side, network, and server-side threats in CORBA systems. Exemplary threats could be "Attackers can tap the network" or "Attackers can send requests to the CORBA server port." After that, the analyst tries to list as many system security vulnerabilities as possible—things like "Sensitive information is transmitted over the network in clear," or "Unprotected IIOP port allows access to server with sensitive information." By doing this task independently from the threat assessment, it is often possible to identify additional threats that previously have been overlooked. Based on the list of threats and the list of vulnerabilities, as well as the value of the endangered information, one can then estimate the corresponding level of risk for each threat/vulnerability pair. If there is a threat without a corresponding vulnerability (or vice versa), then there is no risk.

As part of the risk management step, appropriate countermeasures for each risk are selected and their overall effectiveness is assessed. Sometimes it may be more cost-effective to accept some risks instead of implementing countermeasures if they cost more than the greatest potential loss associated with the risk. In addition to technical and organizational countermeasures, taking out third-party insurance should also be considered, especially for high-impact risks that rarely occur. An appropriate countermeasure for the aforementioned risk examples would be encryption of network traffic and remote party authentication. But even with effective countermeasures in place, there will normally still be some residual risks. The level of these risks and the corresponding countermeasure effectiveness need to be estimated to provide an overview of the remaining security weaknesses of the system. In our network example, traffic flow analysis attacks could be a residual risk, but if the information gained from such an attack cannot be easily used, then the countermeasure effectiveness would be high. However, if the knowledge that a particular party invoked something that gives away critical information, then the effectiveness would be low.

In the final step, the results of the risk analysis have to be documented and signed off on by the management, together with the system description. It is often useful to summarize all parts of the risk analysis in one table that contains columns for the security goals, threats, vulnerabilities, risks, countermeasures, and countermeasure effectiveness. The brief content of each table entry should then be discussed in detail in the remainder of the documentation to make sure everyone understands how the analyst arrived at the results. The details should also contain the value of information, technical aspects of attacks, and evaluations of potential countermeasures.

The effort and expense that goes into developing such a full risk analysis are considerable because it requires a lot of expert knowledge and cannot be automated. In addition, cooperation between all involved parties is necessary to gather all the required information, which further increases the cost. And whenever the system or environment changes, the risk analysis needs to be updated, because any modification can introduce new risks or change existing ones. But despite all the effort, proper risk analysis generally pays off through the knowledge gain that comes with it. Without a similarly systematic approach, acquiring this knowledge is difficult.

2.4.3 Feedback: Analysis and Audit

Maintaining the effectiveness of security policies, guidelines, and procedures in the face of changing systems and environments is just as important as the initial development of adequate policies. From the beginning, clear ownership of the different policy documents (and the responsibilities that come with it) should be assigned. Security procedures can be assigned to security administrators, guidelines to security management, and the information security policy to enterprise management. In addition, an overall management control structure should be put in place to ensure that policies are actually implemented at all levels of the enterprise.

Once the policies are implemented, a team of auditors should analyze the effectiveness of the implemented countermeasures at regular intervals. This is particularly important at the beginning of the system life cycle, as various issues that have been overlooked when the policy's first version was developed may appear at this stage. In addition to overlooked aspects, changes can be triggered by system updates, patches, bug reports, new research results, or changes in the system environment. The feedback provided by the auditors should prompt appropriate modifications to the documents. It is also important to have a well-defined reporting procedure for auditors to prevent policy owners from hiding problems from management.

In Section 2.1, we discussed how countermeasures can be detective, reactive, and preventive. Analyzing policy effectiveness can be carried out in three analogous ways: First, security problems can be identified at the time they occur so that the damage can be restricted. Event management and notification systems such as intrusion detection systems (IDS) can inform the local administrator if a potential attack is taking place, so that appropriate actions can be taken. For example, a network connection can be dropped at the firewall if the administrator suspects that an attack is being carried out over the network.

Security problems can also be identified after they have occurred in the system, so that they can be prevented in the future or help track down attackers. Analyzing security-relevant log files for suspicious actions and security breaches can facilitate this. Tools that filter audit log files and assist in the analysis can considerably reduce the effort.

Finally, security weaknesses can be identified before any security breaches even occur. One rather unstructured approach, called *penetration testing*, involves hiring security professionals who try to attack the system. Unfortunately, the fact that these hackers are unable to attack the system does not automatically mean that no one else can attack it. A more analytical approach is generally preferable, in which security experts perform a conceptual security analysis to identify weaknesses on a more formal basis. Either way, the cost and effort of both these approaches can be significant.

2.5   Threats, Vulnerabilities, and Countermeasures

Three key concepts—threats, vulnerabilities, and countermeasures—come up in most discussions about security. A *threat* is a possible danger to a system that might exploit a vulnerability of the system. A *vulnerability* is a point where a system is susceptible to attack. The more vulnerabilities you can identify in your system, and the more threats you believe are in the system environment, the more carefully you will need to consider how to protect the system and its information. Techniques for protecting the system are called *countermeasures*.

Practical information security is concerned with identifying threats and vulnerabilities to information systems, as well as protecting against threats to those systems by applying appropriate countermeasures.

2.5.1   Threats

A threat is defined as a possible danger to a system; the danger might be a person (e.g., hacker), a thing (e.g., faulty piece of equipment), or an event (e.g., fire or flood) that might exploit a vulnerability of the system. A threat is therefore a potential system misuse that could lead to a failure in achieving the security goals described in Section 2.3. Human attacks are examples of such threats, as are natural disasters, inadvertent human errors, and internal hardware or software flaws.

In general, threats fall into three main categories: environmental, accidental misuse, and intentional attacks. Environmental (i.e., physical) threats are threats that imperil every physical piece of equipment, such as natural disasters, floods, and fires. Accidental threats are normally caused by ignorance or insufficient training of authorized users. Intentional threats can be distinguished into attacks by outsiders, such as hackers or spies, and insiders, such as disgruntled employees. Examples of typical threats in CORBA systems include:

- *Information compromise:* Deliberate or accidental disclosure of confidential data (e.g., masquerading, spoofing, eavesdropping);

- *Integrity violations:* Malicious or negligent modification or destruction of data or system resources (e.g., trapdoor, virus);

- *Denial of service:* Curtailment or removal of system resources from authorized users (e.g., flooding);

- *Repudiation of some action:* Failure to verify the identity of an authorized user and provide a method for recording the fact (e.g., audit modification);

- *Malicious or negligent misuse:* Active or passive bypassing of controls by either authorized or unauthorized users (e.g., browsing, inference, harassment);

In practice, there is no security system that can counter all possible threats to a system, mainly because there are infinitely many (both physical and software-based) interactions with the environment, and the nature of many interactions is unpredictable. For example, although auditors keep administrators in check and vice versa, a group of collaborating administrators and auditors can modify a system as they wish and remove any trace of their actions afterward. Also, some countermeasures to potential threats are outside the scope of what can be countered by any information security system (e.g., natural disasters).

In line with this, CORBA security (and middleware security in general) cannot counter all possible threats to a distributed system—it can only protect resources managed and controlled within the scope of the middleware.

In particular, the threats that should be countered by security measures outside CORBA security are denial of service attacks and traffic analysis attacks, which both should be countered by the underlying communications software, and Trojan horse attacks, which should be countered by a well-defined software change control process.

2.5.2 Vulnerabilities

A vulnerability is a weakness in the security system that might be exploited by a threat. A threat that exploits a vulnerability is said to perpetrate an attack on the system. No matter how many countermeasures you put in place, any system that provides a service to someone will be vulnerable to some residual attacks. Security policies and countermeasures, such as security technology, may reduce the likelihood that an attack will be able to penetrate the system's defenses, or they may require an intruder to invest so much effort and resources that it is just not worth it—but there is no such thing as a completely secure system.

There are various kinds of vulnerabilities, such as physical access to hardware, natural disasters, or human errors. Most of these are beyond the scope of what CORBA security could counter. Examples of typical vulnerabilities in CORBA systems include:

- An authorized user could gain unauthorized access to information that should be hidden from him.

- A user could masquerade as someone else, so that actions are being attributed to the wrong person. In a distributed system, a user may delegate his rights to other objects, so that they can act on his behalf. Therefore, the threat of rights being delegated too widely causes additional problems.

- Security controls could be bypassed.

- Network communications could be subject to eavesdropping.

- Network communications between objects could be tampered with.

- Lack of accountability for (malicious) actions.

Vulnerabilities are often the result of deliberate or unintentional trade-offs made in system design and implementation, usually to achieve increased performance or additional functionality. This is acceptable as long as the risks associated with the trade-offs are identified and documented, and responsibility for this decision is clearly assigned.

2.5.3 Countermeasures

Finally, a countermeasure is a protective measure that reduces or removes a vulnerability. A countermeasure can be an action, device, procedure, or technique. In general, there are many different types of countermeasures, such as computer security, communications security and physical security, as well as policies and procedures.

The CORBA security services implement several countermeasures through the following basic functional components, which will be described in more detail in Chapter 3:

- *Identification and authentication:* Includes authentication of principals and authentication between clients and target objects;

- *Access control:* Preventing unauthorized invocation of operations;

- *Security audit:* Detecting system misuse;

- *Communications protection:* Includes integrity and confidentiality protection based on an authenticated security context;

- *Nonrepudiation:* Generating irrevocable evidence of user actions;

- *Security administration:* Includes policy creation and management;

- *Segregation:* Separating applications from each other, data from functions, user's duties;

- *Automatic security enforcement:* Tries to prevent bypassing of countermeasures during object invocation.

2.6 Middleware Security

Distributed middleware systems such as CORBA are more vulnerable to security breaches than more traditional host-centric systems, because there are more places where the system can be attacked. Thus, CORBA systems have specific security requirements that take into account the inherent complexities that result from their distributed nature.

The prime point of attack is the network, which is often accessible to anyone. Communications can be eavesdropped upon (passive attack), and attackers can masquerade as legitimate participants, because system interactions are often unpredictable and complex (active attack). Hosts are the other main location where distributed systems are vulnerable to various attacks. This is because, in many cases, the middleware and application objects work on top of an insecure operating system that, from an application perspective, cannot be trusted.

In addition to the increased surface area that is susceptible to attacks, there are a number of issues that further complicate security enforcement and administration in distributed systems.

2.6.1   Mutual Distrust

In a large distributed system, some components will not trust others. In a traditional client/server architecture, it is clear who is a client and who is a server, and typically clients trust servers, but not vice versa. In distributed CORBA systems, a single object can be client for one request and server for the next, and therefore the trust relationships are complicated. Although there are many security mechanisms to ensure the identity of a remote component, the system architecture must be designed to ensure that these checks are always performed.

In many distributed middleware environments, the client operating systems or the network cannot be trusted to protect the server's resources from unauthorized access. Untrusted hosts are a particularly major concern for electronic commerce applications where cryptographic keys and payment information have to be stored on the client computer. And even if the client system were secure, the network itself is often still highly accessible.

2.6.2   Dynamic Interactions

In distributed CORBA systems, it is often hard to know exactly who the other party is, because there are many flexible interactions between objects. In many applications, target objects on the network become client objects and call other targets themselves. Some objects can delegate parts of its implementation to other objects, for example, by forking out children objects at run-time that take over some of the work. Firewalls in the communications path make the topology even more complicated.

CORBA objects can be polymorphic, which makes it easy to replace one object by another with the same interfaces. However, this feature facilitates the installation of Trojan horses. Also, because of object subclassing, the implementation of an object may evolve over time, so that interactions between objects in distributed object systems become unpredictable.

In large Internet-like distributed systems, the situation is even worse as new components are constantly being added, deleted, and modified. Furthermore, security policies may be changed at run-time. Such dynamic systems are inherently complex. Consequently, the provision of security is also a complex issue.

From a technical perspective, part of the problem is caused by the fact that CORBA does not provide objects with unique identifiers. Objects are located through their object reference, which changes from time to time, for example, when a server gets rebooted or when an object gets moved to another location. CORBA provides a naming service that can provide objects with identifiers, but most implementations allow the same object to have more than one identifier, or no identifier at all. On the other hand, unique identifiers were originally rejected because they were perceived as an obstacle to scalability.

2.6.3 Scalability

By design, distributed object systems can scale without limit, and security is difficult to enforce in such very large systems (e.g., millions of objects on the Internet). Large, possibly geographically distributed systems are cumbersome to administer, especially if there is no trust between administrators. In large distributed systems, one often comes across multiple differing security policy domains, each one enforcing the security requirements of a part of the system. As a result, security policies must be able to address interactions across policy domain boundaries.

2.6.4 Layers of Abstraction

Because of the transparency provided by middleware layering, a great deal of behind-the-scenes activity is going on, which makes it hard to understand and administer the interactions that take place between the invoked objects. CORBA-style distributed middleware architectures are highly layered, and thus CORBA security is also layered. Complex conversions and abstractions at the layer boundaries are an area in which vulnerabilities can occur, because the meaning of security attributes can become imprecise or even incorrect if it is transformed to fit the architectural requirements of different layers. The complexity of the layering is further complicated in systems where security enforcement is widely distributed, in particular, if differing security mechanisms are in use.

2.7 Summary

CORBA security, and information security in general, is about protecting information assets. Each asset has a value, and protecting it is important because it helps minimize the chances that the value is lost. Protective countermeasures can either detect attacks, prevent compromises in the first place, or react afterwards in order to restrict the damage caused.

There are three central security requirements in any CORBA system, which all cover different aspects of controlling access to information: confidentiality, which is the protection of requests and replies, as well as target operations, from unauthorized reading; integrity, the detection and correction of unwanted changes of transmitted data; and availability, which is about ensuring that authorized users cannot be denied service.

Making users accountable for their actions is an important additional requirement in most environments, as is the general dependability of a system to enforce security in the face of attacks.

Security concerns threats, vulnerabilities, and countermeasures. Threats are potential dangers to a system, whereas a vulnerability denotes a point where a system is susceptible to attacks. Following from that, the index of a threat and a vulnerability is a measure for the risk associated with the danger. Countermeasures, both technical and organizational, are put in place to minimize the risk to an acceptable level.

Distributed CORBA systems are more susceptible to security breaches than host-centric systems because there are more attack points on the network, as well as on the hosts. Also, the systems are often very large and geographically distributed and have several domains with differing security policies. The situation gets even more complicated due to the mutual distrust between objects, the inherent complexity of often unpredictable and dynamic interactions, components being dynamically added and removed, polymorphism and inheritance of objects, and lack of unique object names. Moreover, a great deal of behind-the-scenes activity is taking place under several abstraction layers that exist on each CORBA host. This makes the inner workings of CORBA difficult to understand and can lead to semantic mismatches when security information is passed on from layer to layer.

But the technical aspects alone do not make security in CORBA systems difficult. Managing security effectively is also a hard task. Rather than just focusing on the technical issues, one has to approach security from the enterprise as a whole, including both organizational and technical aspects. Security management is often based on an abstract hierarchy of documents. On the abstract end is the unchanging information security policy, which defines the goals, responsibilities, and commitments regarding security within the enterprise. On a medium level are security guidelines that describe which countermeasures should be implemented. Appropriate countermeasures are identified as part of the risk analysis process. On the detailed end, regularly changing security procedures state in detail how these countermeasures are to be implemented within particular systems. There should also be an audit and feedback loop to make sure that the security remains effective over time.

2.8 Further Reading: Books on Security

When talking about security, it is important to have a clear understanding of the key terminology and concepts. Dieter Gollmann's book on computer security [1] is by far the best and most up-to-date general introduction to technical security. It provides a broad overview with concise definitions and concepts, but at the same time abstracts from unnecessary details. Bob Blakley's book on the CORBA Security model [14] provides a more CORBA-specific, but less solid, introduction to a number of key security concepts. It is short, and written in a light and entertaining style, but not shallow. Although it does not explicitly cover CORBA security, Ross Anderson's book on general security engineering [15] is also worth reading. A nontechnical overview of security can be found in *Computer Security Basics* [16]. It is becoming a little outdated, but it is useful if you are interested in a broad, but not too deep, introduction.

References

[1] Gollmann, D., *Computer Security*, Chichester, U.K.: Wiley, 1999.

[2] Hansen, P. B., *Operating Systems Principles*, Englewood Cliffs, NJ: Prentice-Hall, 1973.

[3] International Organization for Standardization, *ISO IS 15408 Common Criteria Version 2.1*, CCIB, ISO/IEC, 2000.

[4] Bell, D., and L. LaPadula, "Secure Computer Systems: Mathematical Foundations and Model," *MITRE Report MTR 2547*, Version 2, 1973.

[5] Clark, D. R., and D. R. Wilson, "A Comparison of Commercial and Military Computer Security Policies," *Proc. 1987 IEEE Symposium on Security and Privacy*, Oakland, CA, 1987, pp. 184–194.

[6]   U.S. Department of Defense, "DoD Trusted Computer System Evaluation Criteria," (The *Orange Book*), DOD 5200.28-STD, 1985.

[7]   International Organization for Standardization, *ISO 7498-2 Basic Reference Model for Open Systems Interconnection (OSI) Part 2: Security Architecture*, Geneva, Switzerland, 1988.

[8]   OMG, *CORBA Security Service Version 1.7* (Draft Adopted Revision), 2000.

[9]   International Federation for Information Processing (IFIP) WG10.4 (Dependable Computing and Fault Tolerance), http://www.dependabilty.org/wg10.4.

[10] Laprie, J. C., *Basic Concepts and Terminology*, Vienna: Springer-Verlag, 1992.

[11] British Standardization Institute (BSI), *BS7799 (British Standard for Information Security Management, British Standard 7799)*, 1999.

[12] Bundesamt für Sicherheit in der Informationstechnik (BSI), *IT Baseline Protection Manual*, Köln, Germany: Bundesanzeiger-Verlag, 2000.

[13] International Organization for Standardization (ISO), *Information Technology—Security Techniques—Guidelines for IT Security (GMITS)*, ISO/IEC TR 13335, 1996.

[14] Blakley, B., *CORBA Security—An Introduction to Safe Computing with Objects*, Reading, MA: Addison-Wesley, 2000.

[15] Anderson, R., *Security Engineering: A Guide to Building Dependable Distributed Systems*, New York: Wiley, 2001.

[16] Russel, D., and G. T. Gangemi, Sr., *Computer Security Basics*, Sebastopol, CA: O'Reilly, 1992.

3

The CORBA Security Architecture

3.1 Introduction

The main purpose of most security systems is to control access to information, based on a set of security policies. In Chapter 2, you learned how information security can be broken down into a number of different aspects, in particular, confidentiality, integrity, and availability. Accountability and dependability are often additional important security goals.

But these essential security requirements are not the only criteria for designing a useful security architecture. A number of additional considerations also need to be taken into account. Most importantly, the security architecture needs to be designed in such a way that it can be integrated into the system without breaking the functional requirements of the overall application architecture. After all, a security architecture is worthless if it enforces the necessary security policies but, at the same time, renders the system unsuitable for its original purpose. In such systems, users will go to great lengths to circumvent the security enforcement to get the functionality they need to do their jobs. There is a trade-off between making the security architecture as unobtrusive as possible and, at the same time, providing effective security enforcement, and it depends on the particular application where the best trade-off is. For CORBA security, being unobtrusive means that the security architecture has to preserve the design requirements of CORBA's middleware architecture (as described in Chapter 1), in particular, interoperability, transparency, flexibility, portability, and scalability.

This chapter describes the abstract architectural design of the CORBA security architecture. We will start with a general description of a number of design principles (see Section 3.2) and then present the basic functional components of the security architecture (Section 3.3). Note that the effectiveness of the supported functionality will be discussed in more detail in subsequent chapters. For now, we will simply present an abstract architectural model of CORBA security without getting bogged down into too much technical detail or a discussion of its weaknesses.

3.2   Design Goals: CORBA Security Wish List

The CORBA security services specification [1] states that its architecture was designed with a number of goals in mind. These design goals have to be understood as a mission statement (or a wish list) rather than a realistic set of targets, as some of the design criteria are too ambitious and sometimes even conflict with others. For example, there is a fundamental clash between interoperability and flexibility [2], because flexibility involves the customization of functionality, whereas interoperability can only be accomplished through standardized functionality and protocols. Another conflict along these lines is between flexibility and assurance—real assurance can only be certified if the system as a whole is looked at, and when some component is changed, the certification needs to be reconsidered. But flexibility in CORBA security means that components can be changed without affecting any parts on the layers above (this property is also called portability).

The remainder of this section will discuss the main design requirements of the CORBA security architecture. There are also a number of requirements that are less interesting for our conceptual discussion, such as good performance and support for object-orientation. Performance depends heavily on the particular implementation, and it is clear that the CORBA security architecture has to be object-oriented in order to work with CORBA.

3.2.1   Interoperability

The single most important requirement of the CORBA architecture is interoperability across heterogeneous systems. Remember that the OMG was originally founded to establish an architecture that enables interoperability between objects on top of ORBs from different vendors, which in turn run on different operating systems.

Also, as we have already mentioned, the security architecture should aim to integrate with CORBA in a nonobtrusive way, which means that it has to preserve the main CORBA requirements. In order to preserve CORBA interoperability, the security architecture should therefore also be interoperable, which includes several aspects. First, it should be possible to provide a set of consistent security policies across a heterogeneous system in which different vendors provide different ORB and security products. That way, organizations can implement their distributed system without any vendor restrictions and can choose the most appropriate technology. For example, some ORB products are optimized to provide real-time or fault-tolerance properties, whereas others are optimized for speed or small code size. A possible scenario could use a heavy fault-tolerant ORB on the server side and a lightweight ORB product on a wireless client device. Similarly, there are a wide range of CORBA security products that support security technology for different purposes. In some Intranet environments, a full-fledged security system with auditing and nonrepudiation may be required, whereas a simple Secure Sockets Layer (SSL)-based security service may be sufficient for some less critical browser-based applications.

Second, objects that reside on a secured ORB should still be able to interoperate with objects that do not have any security. Of course, such communications will not be secured, and it depends on the particular security requirements of the application (on the security-enabled end) if this is advisable or not. From an architectural viewpoint, this requirement means that the security protocols have to be layered over the unsecured interoperability protocols, and that the security enforcement has to be integrated into the ORB communications path in such a way that it can be switched on or off, depending on the security policy for each invocation. There are plenty of possible application scenarios in which an object calls numerous other objects with varying levels of security. For example, an electronic shopping cart application calls a catalog object without any security features to allow the client user to browse the products on sale. Once the user has made a selection of goods, the application would then call the payment object to carry out the purchase—and, of course, this call would need to be secured.

A third aspect is interoperability across domains that support different security policies (e.g., different access control attributes). This feature can be useful in large-scale systems in which callers select their targets dynamically and therefore cannot agree a priori on a common set of attributes. The only viable way to achieve this would involve converters that map attributes from one security policy to the other, but even that may sometimes be impossible if the semantics of the attributes differ dramatically. The current version of the CORBA security specification therefore explicitly excludes interfaces and protocols for cross-domain interoperability.

Finally, it may appear to be useful to support interoperability across systems that support different security technology (e.g., different authentication mechanisms). The advantage of cross-mechanism interoperability would be that an appropriate set of security mechanisms could be chosen for each application environment without inhibiting interoperability. For example, the Secure European System for Applications in a Multivendor Environment (SESAME) would be too heavy to support on a wireless device, whereas SSL may be too limited in its functionality to protect corporate Intranet applications. But when the wireless device is used to connect to the Intranet, secure interoperability should still be possible.

Of course, there are various general problems with this scenario and with cross-mechanism interoperability. First, the representation of security policy attributes in CORBA (e.g., for authentication) is not fully mechanism-independent, which means that the policy can only be evaluated if the corresponding mechanism is supported. But even if the attributes were somehow mechanism-independent, there would be a number of semantic problems: An authenticated SSL identity does not always specify the same object as a Kerberos identity, so it is unclear which one is meant by the abstract identity in the policy [3]. In addition, the fact that a particular mechanism is used is implicitly part of the policy. For example, a target object that services callers that support a stronger authentication mechanism may want to reject callers that have been authenticated by weaker mechanisms. Therefore, it may not be desirable to abstract from the underlying security mechanism at all. On the other hand, not abstracting means that CORBA security will not be able to interoperate if the underlying security mechanisms do not match.

Without any abstraction, cross-mechanism interoperability is difficult to achieve, in particular if incompatible cryptography is used. As previously described for interoperability of policy attributes, it is possible to design converters that map invocations from one technology to another. But this way, the converter has to become a trusted third party with access to all cryptographic keys, which breaks end-to-end security, in particular peer authentication. Due to all these problems, the CORBA security specification should not be expected to ever support interoperability between security mechanisms.

3.2.2 Transparency and Abstraction

Another big advantage of CORBA is that it insulates application objects from lower-level details and complexities—the middleware layer abstracts technical details, such as locating remote objects and mediating invocations to them. The fact that CORBA is largely transparent on the application layer simplifies distributed applications programming, and therefore reduces the overall application development cost. In addition, application programmers do not have to be distributed systems experts to connect their application components.

The security model has to fit seamlessly into the overall middleware architecture to preserve these transparency requirements. As a result, application development should ideally be insulated from security policies and enforcement, just like all the other technical details that happen behind the scenes in CORBA-based systems. Ideally, the security system is simply inserted into the communications path, so that it can enforce the security policy whenever an invocation arrives at the ORB (either from the application above or from the underlying network). The security policies also reside on the middleware layer and specify the required protection (e.g., which invocations are to be passed on or if their occurrence should be logged).

Despite the general requirement that application developers are not required to know about security policies and enforce them for their applications, it should optionally be possible to enforce application-specific security for security-aware applications. As a result, the CORBA security specification distinguishes between *security-unaware* applications that have been developed without any concern about security, and *security-aware* applications that have been designed with application-specific security features in mind. Security-unaware applications are protected transparently by various *ORB layer security features*, whereas security-aware applications can access a range of *application layer security features* to enforce more application-specific and fine-grained policies. Full transparency is, of course, only a requirement for ORB layer security features.

But the security architecture should not just cater to application programmers. From an end-user perspective, security should also be as transparent as possible. In particular, the architecture should support single sign-on, so that users can log onto their computer once and then reuse the generated credentials for a number of applications until they log out or the credentials expire. This single sign-on feature allows application developers to write applications in which security enforcement is almost entirely transparent to users. Also, application development can be done without concern for security, as basic security policies can be enforced on the middleware layer (i.e., transparent to the applications).

Administrators have to carry the main burden for middleware security, as they have to set sensible security policies for all involved components. To do their job, administrators need to look behind the scenes of the middleware layer to obtain the necessary in-depth knowledge about the functionality and location of different application components. The administrative model should be simple to understand and manage, and should provide a single consolidated view of the system. It should allow flexible and fine-grained security policies and, at the same time, support clustering of users and objects (e.g., in roles and domains) for scalability. Most issues of the specified CORBA security model become evident for the administrator—a number of security attributes in policies contain mechanism-specific information that breaks the single view of the system, as well as the abstraction from the underlying technology; moreover, some parts of the model are based on inappropriate security information such as the object interface type. We will discuss these problems in more detail and propose alternative solutions where possible (in Chapter 6).

3.2.3   Flexibility, Portability, and Integration

CORBA was designed to be flexible enough to work in a variety of different environments, ranging from relatively static Intranets with integrated legacy applications on the one end, to wireless devices on the dynamically changing Internet on the other. Such diverse CORBA objects can communicate if there is an IDL language mapping for all participating platforms and if all nodes communicate through the same CORBA protocol (e.g., the GIOP). Most other aspects of a CORBA system, such as the underlying transport layer, can ideally be flexibly replaced without affecting interoperability.

   In line with this, the security architecture should also be flexible enough to fit a wide range of applications and environments with differing security requirements. The model should allow a variety of different security policies and security features, depending on the level of protection required for information on the system. To accomplish that, the model's security attributes and attribute groups can be extended ad lib to reflect additional requirements. However, the use of extended attributes will inhibit interoperability, as they only work with security implementations that support the same extended attributes.

   In addition, the CORBA security model should be independent of the underlying security technology.[1] For example, interfaces specified for security of client-target object invocations should hide the used security mechanisms from both the application objects and the ORB (except for some security administrative functions). The CORBA security model is segregated from underlying security mechanisms that are only accessible through a standardized interface (based on GSS-API [4]). The specification illustrates how security mechanisms are integrated for a number of standard mechanisms (e.g., Kerberos5 [5–6], SESAME [7–8], SPKM [4], SSL/Transport Layer Security (TLS) [9]).

As you have already learned, interoperability problems are caused if different CORBA nodes use differing security mechanisms. In fact, there is a fundamental clash between interoperability and flexibility [2], because flexibility involves the customization of functionality, whereas interoperability can only be accomplished through standardized functionality and protocols. On the other hand, the security architecture allows the replacement of underlying security mechanisms as long as they are replaced on all participating nodes at the same time. But there are implications for middleware layer security policies and attributes. For example, authenticated identities in security policies are mechanism-specific (e.g., an X.509 certificate), which means that the content of the attribute also needs to be replaced when the authentication mechanism changes. The same applies to message protection—the model should support both symmetric and asymmetric cryptography, but all keys will have to be changed when the mechanism is replaced.

To allow application objects to be ported to domains that enforce different security policies and use different security mechanisms, security-unaware applications should be independent from the underlying middleware security system. This way, the underlying technology and policies can be replaced without affecting the application code. Note that application portability without modification is only possible for security-unaware applications, as the security is enforced and administered entirely below the application. If an object enforces application-layer security, then the interfaces to the CORBA security services should hide the particular security mechanisms used (e.g., for authentication). Note that successful porting of applications that enforce their own security to systems with different underlying CORBA security services and mechanisms depends on the attributes used to describe the application layer policy. In most cases, policies have to be changed to reflect the new underlying technology.

But in many application scenarios, CORBA security does not operate on its own. Often, a security infrastructure of some sort is already in place

---

1. In a number of countries, the export of cryptographic software is regulated as part of dual-use export regulations. The security model has to take this into account and make it possible to ship a CORBA security services implementation without the actual encryption part. The security service can then be integrated afterwards with a local cryptographic package.

that has to be integrated into the CORBA system. Therefore, if the system already provides security protocols and mechanisms, it should be possible to reuse these without the need for new cryptosystems, logons, access control repositories, user registries, or policy databases. To achieve that, the architecture is specified to be mostly independent from the underlying security mechanism and, thus, can integrate with a wide range of security mechanisms and environments, as long as the mechanisms provide sufficiently well-specified interfaces. In practice, this is only possible if the source code of the CORBA security service implementation is available, so that preexisting mechanisms can be integrated into the security architecture.[2] The specified model supports a number of different security policy types for access control and audit, and integrates with various different security mechanisms, which can be reused if they are already in place in the particular environment. However, the model does not provide mechanism-independent security policy attributes, in particular for access control and audit.

In addition to integrating preexisting security mechanisms, the model should allow the provision of consistent security policies across heterogeneous systems that contain legacy applications. The middleware architecture can enforce security for legacy system components by putting so-called CORBA wrappers around them, which allow the provision of consistent security policies, as long as the CORBA security systems (and the underlying security technology) match on all participating nodes.

3.2.4   Scalability

The security model should support CORBA systems of different size, ranging from small to very large. The security model as such does not impose any size restrictions; potential upper limits to the number of participants or policy entries are purely implementation-specific. The size limit for secure CORBA systems is often restricted by underlying security mechanisms' scalability restrictions and policy implementation.

But the CORBA security architecture should provide the means to make administration of large-scale secure systems easier. To reduce administrative overhead, individual identities should be grouped into *roles* (or groups) with the same privileges. Analogously, objects that share the same security policy should be grouped into *domains*. Note that despite its support

---

2. The source code of both MICO, the example of ORB used throughout this book, and its MICOSec security services implementation, is freely available under the GNU public license (see Section 4.2).

for roles and domains, the current architecture does not include any tools to manage them in an effective and interoperable way. The OMG is currently working on an additional security specification to manage domain membership [10] in order to solve this problem.

There is also a requirement to manage the distribution of cryptographic keys securely and with low administrative overhead. To provide key management facilities, the OMG currently specifies how public key infrastructures (PKI) should be integrated into CORBA security [11]. Again, the upper limit depends on the actual PKI implementation.

3.2.5 Reliability and Assurance

Assurance is a qualitative measure of trustworthiness [1]. A security architecture is trustworthy only if all applicable security policies are enforced on all actions. In the context of CORBA, this means that it should not be possible for malicious principals to bypass the security system and invoke a method on the target without triggering the security policy enforcement and protection required by the security policy. The policy might state in some cases that no security functionality should be applied, but this has to be determined as part of the security policy evaluation process, and so the security system is not bypassed after all.

CORBA provides standard ORB interfaces at which the CORBA request traffic can be intercepted as it gets passed up or down through the ORB message path. These so-called interceptors are a convenient place to integrate the security system with the ORB. However, this means that not just the security service implementation and all underlying security mechanisms have to be trusted to work reliably but, in fact, the whole ORB also has to be trusted. The U.S. *Trusted Computer System Evaluation Criteria (TCSEC)* [12] calls the set of all system components that has to be trusted the trusted computing base (TCB) and suggests that security-relevant components be segregated into a (preferably small and well-understood) security kernel. The underlying idea is that the system's overall security is only as high as the security of the weakest component in the TCB.

Unfortunately, this monolithic TCB approach is not suitable for many CORBA environments, partly because of the complexity of distributed middleware systems, and partly because secure interoperability of CORBA applications and ORBs is often based on mutual suspicion. Therefore, the specification defines the concept of a distributed TCB, which denotes the collection of objects and mechanisms that must be trusted so that end-to-end security between client and target object is maintained.

The usefulness of such a definition for real-world systems is questionable, in particular since it is impossible to put this large number of TCB components into a small and well-understood security kernel. A distributed TCB in CORBA can potentially contain the whole stack below the application object: the ORBs and object adapters, the associated ORB services, the security services, underlying security technology, the supporting operating system, and the lower layer communications software (see Figure 3.1). The distributed TCB does not extend across several nodes if there is mutual suspicion between the caller and target. But this is normally not a problem as it is infeasible for attackers to bypass the remote TCB from the network (unless someone hacks into the underlying operating system). For example, even a malicious client with a rogue application program, ORB, and security service cannot invoke methods on any target without triggering the target-side security enforcement—the application object does not listen to any network ports, it is the ORB that will pick up any requests. Therefore, any irregularities in the authentication or message protection mechanisms (e.g., disabled encryption) would be noticed by the target security service and the request would not be passed up into the application.

But trust is also highly related to reliability, which is more an implementation issue than an architectural concern. Any CORBA security system may become unreliable if software bugs in the TCB can be exploited and, due to the high degree of complexity, bugs are almost unavoidable. But, of course, this applies to any complex system, not just to CORBA and CORBA security. For example, one of the authors (as an independent security

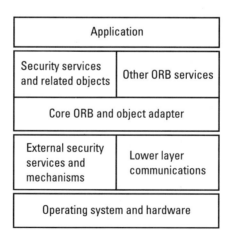

Figure 3.1 Components of the distributed TCB.

auditor) once encountered a CORBA security system that would encode a signed session ID into the transmitted byte stream to allow the receiving side to associate the message with a particular session. Messages without ID would be transmitted unencrypted. One day, the problem occurred that, due to inverse byte ordering on differing underlying hardware and software platforms, neither side could find any session IDs and assumed that all communications should be unencrypted. All testing had been done in a homogeneous environment where this bug did not manifest. It was assumed by the developers that the underlying Java Virtual Machine (in line with its specification) would abstract from these underlying differences. In summary, the resulting system would transmit sensitive information unencrypted for CORBA communications across differing underlying hardware, just because of a bug in one of the underlying layers.

The certification process approves the trustworthiness of a system to a certain degree and, as a result, establishes confidence in the implemented security measures. According to the specification, the security model should support accreditation as defined in government criteria, such as the European *Information Technology Security Evaluation Criteria (ITSEC)* [13] and the *Common Criteria* [14]. However, in practice, it depends on the quality of the implementation process and the style and quality of the implementation, rather than on the CORBA security model itself.

3.2.6 Simplicity

To most readers' surprise, the 420-page CORBA security services specification [1] states in the beginning that the CORBA security model should be easy to understand and administer, which implies that it should only have a few concepts and components. Following our discussion about the desire to have a trusted computing base that defines a small and well-understood security kernel, it is clear that a simple model would be likely to result in a more trustworthy system. A simple model is normally easier to implement; therefore, the scope for bugs and misinterpretations is limited.

However, as the mere page number of the specification shows, the CORBA security architecture does not achieve this target. This is mainly because it was designed to incorporate a large number of features for a number of differing environments and scenarios (e.g., automatic security enforcement for security unaware applications but also fine-grained security functionality for security-aware applications). As a logical consequence of that, and because of the inherent complexity of distributed applications security, the resulting architecture is bound to be complex as well.

In Section 3.3, you will get a feel for the complexity of the security model, even though many tricky details are hidden at this highly conceptual level. It all makes sense when a number of boxes are drawn and the flow of data between them is described informally, but many complications only become obvious when the model is actually implemented in practice. You will learn more about intricate technical difficulties of practical CORBA security in Chapters 5 through 7.

3.3 Architectural Components

The CORBA security architecture consists of a number of functional components that provide applications with security. This section describes the security models of the core components for authentication, session establishment, delegation, policies and domains, access control, message protection, security audit, and nonrepudiation. The purpose of the diagrams presented is to illustrate on an abstract level how the main objects for each functional component interact. Note that they do not always reflect the exact object interfaces and data flows, but are rather meant to give a conceptual understanding of the model as a whole.

In principle, the following aspects are covered for each functional component: First, the corresponding security policy will be discussed; then the policy's evaluation process is examined; and, finally, the policy's enforcement is illustrated together with its possible location in the architecture (ORB layer or application layer).

Despite the fact that the model is logically separated into various functional components, it is important to appreciate the many interdependencies between them. For example, if the access control component uses the peer identity as an attribute, then it relies critically on the functionality provided by the authentication component.

3.3.1 Principals and Credentials

Before talking about the functional components of the security model, it is necessary to identify who is going to populate it. In classical security models such as Bell-LaPadula [15], there are two groups of entities in a system—subjects and objects. Users want to access resources, and the security system has to make sure that access to a resource is restricted to authorized users. Other models such as Clark-Wilson [16] introduce another entity called procedure, which controls access to resources and can only be called by certain users. In such models, it is assumed that users cannot access resources directly.

The CORBA security model is based on active subjects who can invoke operations on target objects, but only by going through a security enforcement component. Subjects and objects can have an identity and a number of privileges (or credentials), which the security enforcement is based on.

3.3.1.1 Principals

A *subject* in CORBA security is an active entity in the system that tries to use the system or its resources. A subject is assumed to have an independent will, which makes it different from other system entities that only react to invocations. Subjects can be human users, but nonhuman subjects also exist. Programs can be nonhuman subjects if they carry out actions without specific commands from users, such as hardware devices or active software daemons.

In the security model, active entities are called *principals*, and one of the central properties of a principal is its *identity*. An identity describes its principal uniquely in the sense that no two principals' identities may be the same. But at the same time, principals may have several different kinds of unique identities. For example, a principal may have both a unique audit identity and a unique access identity.

It may not be clear why not only human users, but also software components should be principals in the sense defined above. After all, they are always started by a user (or administrator) and could therefore—as in the operating systems world—inherit the identity of that user. The reason is that objects in complex CORBA systems often need to be authenticated by their own separate identity and not by the identity of the administrator who started it. An administrator might, for example, start a whole range of services of differing quality (e.g., response time, bandwidth), and the caller wants to make sure it is actually connecting to the type of service for which it is paying. This would be impossible if all these services had the same identity inherited from the administrator. To achieve this, each component has to authenticate itself to the CORBA security services with some stored authentication information, such as an identity certificate.

3.3.1.2 Credentials

The CORBA security system has to decide what it will do with each principal's actions. The information used for this decision is stored in the security policy. To enforce the policy, the system needs to know for sure who the principal is, but it also needs to be able to describe the security-relevant properties for all principals, which are called *credentials.*[3]

The goal of the CORBA authentication process is the generation of such credentials.[4] Credentials are the information that describes the security attributes of a principal [1], whereby attributes can be identities or privileges of the principal (or both). A *privilege* is a security attribute that, as opposed to an identity, does not need to be uniquely associated with a principal. Examples of privileges include groups, roles, and clearances. An identity could be the name of the user, whereas a role could be his or her job title (e.g., administrator). The security attributes described in the credentials express the principal's characteristics, which form the basis of the system's policies governing that subject.

Whenever the CORBA security system encounters a new and unknown subject, it automatically assigns a default credential to it, which contains no identity and only one privilege attribute, called "public." For such subjects, a default policy will be enforced. For some subjects, this may be sufficient. Other subjects require more (nonpublic) privileges, which can be done by authenticating to the security system (see Section 3.3.4). After establishing the authentic identity, the security system will assign additional attributes to it, based on the security policy (see Figure 3.2).

3.3.2 Administration: Policies and Domains

If all objects and principals in large-scale distributed systems were administered individually, security management would become more and more cumbersome the larger the systems get. Therefore, objects that have common security requirements are grouped into domains. A domain is a distinct

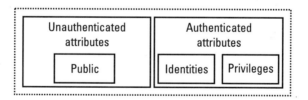

Figure 3.2 Credentials and security attributes.

---

3. Note that not only clients can have credentials. Targets can also have credentials that contain their identity attributes, etc.

4. Credentials can also be obtained through delegation, which will be described in Section 3.3.3.

scope, within which certain common characteristics (policies) are exhibited and common rules are observed (see Section 3.2.4). Domains can bring the security administration down to a manageable size.

CORBA security policies specify the level of protection required in the domain, as well as the actions to be taken for each event. The result of the policy evaluation depends on a number of security attributes, such as the caller's authenticated identity and its other credentials. Domains can be different for different security policies (e.g., access control domain, audit domain). The ORB makes sure that the policies that apply within a domain are automatically enforced for each object in that particular domain. In practice, domains could be centrally managed in a domain server that contains the policies for its domain (e.g., access control or audit domain server).

Note that cryptographic keys may be necessary for clients to authenticate themselves to the domain server and to protect sensitive credentials from unauthorized access. These keys need to be set up for all clients, and there must be keys for secure communication between domain servers from different domains. Such key infrastructure support is currently not specified for CORBA, but there are some current efforts to integrate PKIs [11].

Several types of domains are relevant to security: security policy domains, security environment domains, and security technology domains. On a lower level of abstraction, protection domains are used to cluster objects that trust each other.

3.3.2.1 Security Policy Domains and Roles

Security policies are the rules and criteria that specify which security features have to be enforced to protect objects and resources. CORBA security policies define the rules for authentication, secure invocation, privilege delegation, access control, audit, and nonrepudiation [1]. Keeping the policies independent from the application code allows security administration for security-unaware objects without changing the application code.

Domains

A *security policy domain* clusters all objects to which a common security policy applies (see Figure 3.3). This way, it is possible to administer many objects with only a few policies, which helps deal with the problem of scale of distributed object systems. In addition, administrators' authorities can be separated by delimiting the scope of administrative activities to one domain (this is often called "separation of duties").

When an object is created, it automatically becomes a member of one or more domains and, therefore, is subject to the security policies of those

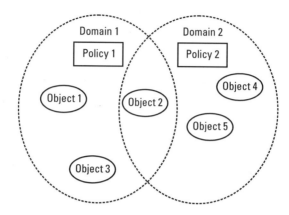

Figure 3.3 Security policy domains.

domains. In other words, a security policy domain is the scope over which a common security policy is enforced. A domain can have subdomains that reflect organizational subdivisions (e.g., departments).

In CORBA, it is possible to have overlapping policy domains, either of the same type or of different types (e.g., access control policy, audit policy). Also, domains can be federated, whereby each domain keeps most of its authority but agrees to give members of other domains limited rights. A federation must be able to handle policy differences across domains (e.g., mapping of access roles across domains).

Table 3.1 illustrates conceptually how such a policy table could look for a particular domain "Bank".

Roles

So far, we have grouped objects into security policy domains. In larger systems, it also makes sense to cluster principals into roles (or groups) to reduce administrative overhead (see Section 3.2.4 and Figure 3.4). This way, the policy does not need to state how every individual identity has to be treated.

Table 3.1
Access Control Policy Example

Policy for domain "Bank"		
Caller identity = "Rudolf"	Invoked operation = "balance"	GRANT
Caller identity = "Ulrich"	Invoked operation = "withdraw"	GRANT
*	*	DENY

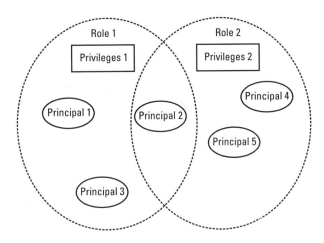

Figure 3.4 Roles.

Instead, it uses roles such as "administrator" or "auditor." Any principal that owns such a role can carry out the actions associated with that role. Note that a role is not an identity, as it is shared among a number of principals. In particular, the audit and nonrepudiation functions often require a unique identity for each principal to be able to associate responsibilities for actions to individuals.

ORB Layer and Application Layer Security Policy

Security policies can be enforced either by the ORB or by the application. The ORB layer security policy is enforced by the ORB (and by the security services it used, as well as the underlying operating system (OS) that supports it). Policies of this type are transparently enforced for both security-aware and security-unaware objects.

Application security policies are enforced by security-aware application objects, which may have their own security requirements. Application security policies may either be grouped in domains or individually customized to a certain application.

Note that the contents of security policies are mostly domain-specific (e.g., specific to the used security technology), which often makes cross-domain communication impossible. To a certain extent, the specification tries to standardize the semantics of security information (e.g., access rights—see Section 3.3.6 for more details) to mitigate this problem. However, there are some fundamental problems with interoperability when policies should be flexible enough to fit a variety of different application domains [2].

3.3.2.2 Other Domains

In addition to security policy domains, the specification defines some other, more conceptual, domains. These are security environment domains, security technology domains, and protection domains. Grouping principals and objects into these domains should make it easier to deploy and manage secure CORBA systems.

Security environment domains specify the scope over which policy enforcement may be achieved by some means local to that environment (i.e., not by the CORBA security services). For example, extra encryption may not be necessary when messages are being transferred between objects on the same machine, because there is no untrusted network. Environment domains should be exploited where possible, to optimize performance and resource use, since enforcement of one or more policies (and their associated mechanisms) is not needed. Two possible types of environment domains are considered in the specification: message protection domains, where integrity and confidentiality protection are available by some specific means; and identity domains, where objects can share the same identity.

Security technology domains are domains in which common security mechanisms are used to enforce policies. For example, the same technology is available for principal authentication and the same authentication services are used, or data in transit is protected in the same way, using common key distribution technology with identical algorithms. The purpose of security technology domains is to identify which objects use the same underlying security services. Distinguishing this type of domain helps set up and maintain the domain's underlying services and administer entities in the way required by this security technology. Also, security technology domains can help identify where bridges and gateways are necessary for interoperability between security technology domains. Note that the specification does not include any security technology-specific administration interfaces.

Protection domains cluster components are assumed to trust each other. The security architecture should be designed to make sure (at least in theory) that components from different protection domains cannot interfere with each other. Maintaining integrity and confidentiality in a secure object system depends on proper segregation of objects, which may include separating security services from other components. It must be possible to guarantee that, to any required level of assurance, the applications cannot bypass them. Moreover, security services themselves must also be subject to security policies. The general approach is to establish protection boundaries around groups of components that are said to belong to a protection domain.

Components belonging to a protection domain are assumed to trust each other and no security is needed, but communication across protection boundaries has to be controlled. It must be possible to constrain interaction between components to controlled communication paths, for example, explicit message passing and implicit sharing of memory.

3.3.2.3 Security Administration

The purpose of security policy domains and roles is to make policy administration of large systems easier. In addition to domain management, security administration also involves setting up domain servers if required, policy management, interdomain interoperability and policy agreement between domains, security mechanism installation and management, providing secure storage for audit trails and keys, and security service set-up. In general, security administration tries to bring together all the different components in such a way that they provide a secure system without any loopholes (as opposed to just individual security-enforcing components).

In this section, we will only discuss policy and domain administration. The CORBA security service defines a `DomainManager` interface that allows policy objects to be created, deleted, and updated for different security functions. Interfaces are provided to locate `DomainManagers`, but management of policy domains and their members is currently not supported. Note that there is a submission for a *Security Domain Membership Management Service* (SDMM) [10], which maps objects to domains. This way, security enforcement can be based on unchanging domain names rather than the less useful interface type or the often unpredictable object reference.

Managing security environment domains and security technology domains is often done in an environment-specific and security technology-specific way; thus no interfaces are specified in the CORBA security service. The specification also does not cover administrative functions concerning the management of underlying mechanisms supporting the security services, such as authentication services, key distribution services, or certification authorities. Note that this raises some chicken and egg problems that have to be carefully considered.

3.3.2.4 Interoperability

Secure interoperability between objects from different security domains is often only possible if they both happen to use the same security attributes and mechanisms, or if a gateway translates object references and messages between domains. It is not a goal of CORBA to specify such security gateways (to translate security mechanisms) and bridges (to translate policies), as their functionality is highly implementation-specific.

Instead, the specification tries to mitigate the problem by specifying tokens for some security mechanisms (e.g., Kerberos) and some standard security attributes (e.g., standard access rights), which have to be supported by all compliant implementations. This way, objects can at least interoperate across domain boundaries with a minimum level of protection.

3.3.3   Privilege Delegation

In distributed object systems like CORBA, an invoked object may, in turn, call other objects to perform parts of the task. The resulting chain of calls complicates the credentials model (see Section 3.3.1), as credentials will need to be passed from one object to the other, and intermediate objects will enforce their own policy on these credentials before they pass them on to the next target. For example, access decisions may need to be made at each point in the chain, whereby intermediate objects may use different authorization schemes and may, therefore, require different access control information to check which objects in the chain are permitted to invoke further operation on other objects.

This process is called privilege delegation—the act whereby one principal authorizes another to use its identity or privileges, perhaps with restrictions. The owner of the original credential is called the *initiator* of the request, and the recipient of a delegated request is called an *intermediate*. If an intermediate chooses to turn the initiator's received credential into its invocation credential, then it becomes a *delegate*. The final recipient of the request (i.e., the object that performs the requested operation) is called the *target* of the request.

In privilege delegation, the initiating principal's security attributes may be delegated to further objects in the chain to give the recipient the right to act on its behalf. Intermediate objects may need to use their own credentials for some operations and delegated ones for others. The delegation model allows the initiator to restrict delegation of some of its security attributes. If no restrictions are placed and only the initiator's privileges are being used, this is called impersonation.

3.3.3.1 Policy

There are delegation policies for initiators, all intermediates, and targets, which reflect the interests of all involved parties.

- The initiator can decide whether or not to allow its credentials to be delegated (by setting the `DelegationMode` either to `Delegate` or to `NoDelegate`). This is an important decision because it is often not easy to revoke delegated credentials once they have been given away.

- Each intermediate's policy can specify (on a per-interface granularity) if it should pass on the initiator's credentials (simple delegation), its own credentials (no delegation), or a combination of them (composite delegation, combined privileges delegation, traced delegation). Figure 3.5 shows the different delegation schemes supported by the CORBA security model.

- The target policy specifies which requests should be accepted, based on the way the associated credentials have been delegated. For example, an invocation could only be granted if composite delegation has been applied, so that the whole delegation chain can be evaluated.

3.3.3.2 Enforcement

Controls can be enforced on the client side (including initiators and intermediates) before initiating object invocations. Interfaces allow control of privileges delegated, control of target restrictions, and control of time restrictions, which means that a client can control the delegation of its credentials by specifying a time window during which the delegated privileges are valid. The privileges will automatically expire outside this time interval. The client can also specify the maximum number of method invocations for which a delegated credential is valid.

The intermediate object is able to extract received privileges from the request and the active security context, and use them in local security decisions, or when making the next invocation. It can also build (if permitted) new credentials with changed attributes, using the received ones.

The target object uses the received privileges for local security decisions when the delegated request arrives, for example, for access control. Note that all delegation policies are, in fact, enforced by each receiving side, and ultimately of the target object that carries out the operation intended by the initiator. All the client can do is protect the integrity of its delegation credentials so that no intermediate can modify them, but it cannot control what intermediates do with these credentials. For example, a malicious intermediate could pass on credentials even if the `NoDelegate` option is selected. It is

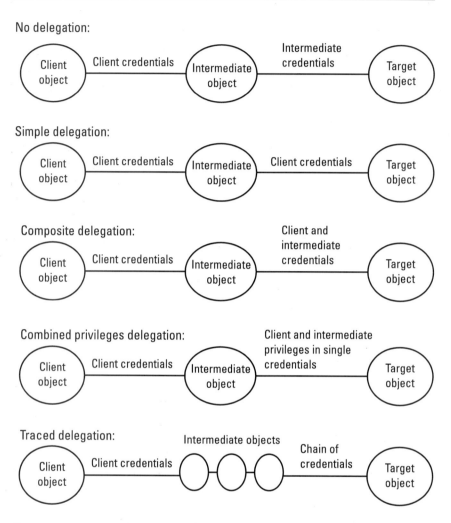

Figure 3.5 Privilege delegation schemes.

then up to the target to find out during authentication that the credentials do not belong to the intermediate and, consequently, reject the request.

3.3.3.3 ORB Layer and Application Layer

Of course, only security-aware applications can select delegation schemes and specify target restrictions for themselves. For security-unaware applications, the administrator has to specify which delegation policies should be used by default when an object acts as an intermediate. This allows many applications to be unaware of the delegation options used, as many controls are done automatically by the ORB when the intermediate invokes the next object in the chain. The difference between security-aware and security-unaware intermediate objects is illustrated in Figure 3.6.

Note that CORBA security reuses delegation functionality of the underlying security mechanism and is, therefore, limited to the functionality provided by the security mechanism (which is, in most cases, inadequate or nonexistent). However, the *Common Secure Interoperability (CSI) Version 2 Specification* [17] solves this problem by introducing an additional protocol layer that supports the use of delegation tokens.

3.3.4 Principal Authentication

The CORBA security model divides the process of discovering who a principal is into three basic steps. First, the system asks a principal to identify itself. Then it authenticates the claimed identity, because the principal might be lying. Finally, after establishing that the identity is authentic, it checks the privileges of the principal within the system.

In other words, authentication is the process of verifying a claimant's claimed identity, more specifically "The verification of a claimant's entitlement to use a claimed identity and/or privilege set" [1]. Different types of authentication information, such as passwords or cryptographic keys, are used to establish a claimant's entitlement to a claimed identity. The critical point here is that the authentication information is assumed to be known only to the entity associated with the identity. In the case of passwords, this means that a claimant has to know the corresponding password during login to verify the claimed username. In the case of cryptographic keys, the claimant has to provide a valid certificate (signed by a certification authority) that binds his cryptographic key to his identity. The cryptographic certificate validation mechanism then ensures that this binding is valid.

The authentication process bootstraps the entire security system. All other components rely, in one way or another, on the privileges verified during the authentication process. For example, access control is often based on the caller's identity, which relies on successful authentication. The same applies to audit, where logged events normally have to be associated with the event initiator to provide accountability.

Note that this section is only concerned with the authentication of principals to the system. Peer authentication for remote communications is covered as part of the secure context establishment (see Section 3.3.5). In the CORBA security model, peer authentication relies on the credentials

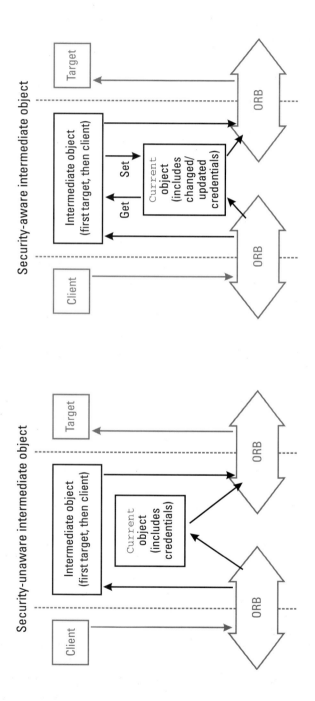

Figure 3.6 Intermediate objects.

generated on both sides during the principal authentication process. This way, a principal has to log in only once to establish a number of authenticated security contexts.

3.3.4.1 Principal Authenticator

The central object of the authentication model is the `PrincipalAuthenticator`, which provides an `authenticate` method for principals to authenticate themselves and create their credentials. In addition to supplying its claimed identity and associated authentication information (e.g., a password or certificate), the caller can specify the authentication method to use (e.g., password validation), and the security mechanism with which to create the credentials (e.g., X.509 certificate). If the principal wishes to use only a subset of its privileges in a particular session, then it can also request specific privilege attributes.

The `authenticate` method then creates specific credentials for the caller, depending on the information provided. The newly created credentials object is then placed into the `Current/SecurityManager` object so the credentials can be used during security session establishment (see Section 3.3.5). It can also return mechanism-specific data and some continuation data if authentication proceeds in several steps (e.g., for challenge/response authentication mechanisms). This continuation data is used in subsequent steps of the authentication protocol. Figure 3.7 illustrates the inputs and outputs of the `PrincipalAuthenticator::authenticate` method.

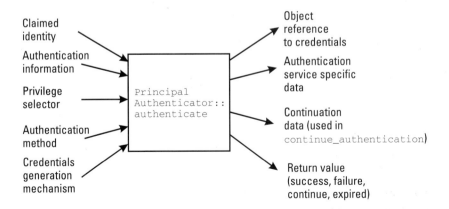

Claimed identity

Authentication information

Privilege selector

Authentication method

Credentials generation mechanism

Principal Authenticator:: authenticate

Object reference to credentials

Authentication service specific data

Continuation data (used in continue_authentication)

Return value (success, failure, continue, expired)

Figure 3.7 `PrincipalAuthenticator::authenticate.`

3.3.4.2   User Sponsor

Human users provide their claimed identity and authentication information through a component called the *user sponsor*. The user sponsor then calls the `PrincipalAuthenticator` on behalf of the user to generate the credentials. In the simplest case, the user sponsor is just a login window that asks the user to enter his or her login and password. More elaborate user sponsors can support more complex authentication mechanisms, such as authentication based on physical tokens or biometrics.

The specification does not define any interface for the user sponsor, because it may reside either inside or outside the CORBA system. For example, the user sponsor could be part of a preexisting authentication infrastructure, such as an operating system logon. This important feature allows many secure CORBA and non-CORBA applications to share the same logon, which is often referred to as *single sign-on* (i.e., a user only has to log on once for each session and not for each individual application). In most cases, the user sponsor is only useful on the client side because the target wants to know which individual user is behind the client application. On the target side, nonhuman principals (application objects) authenticate themselves by calling the `PrincipalAuthenticator::authenticate` method directly.

3.3.4.3   Policy

The principal authentication policy specifies which identities and privileges are to be given to a principal based on the presented authentication information. This policy is not explicitly modeled in CORBA security, but is rather implicitly enforced by the underlying authentication mechanism. Based on the provided security information, the authentication mechanism will decide which privileges to put into a principal's credentials object.

Note that the generated credentials will contain some authentication mechanism-specific data, such as an X.509 certificate that identifies the principal. Because of that, the CORBA security model can only support centralized administration of principal authentication policies with considerable effort.

3.3.4.4   Enforcement

Principal authentication policies are enforced by the `PrincipalAuthenticator` and, ultimately, by the underlying authentication mechanism. The model only describes how the underlying authentication mechanisms are integrated into the security architecture. For example, if Kerberos is used, there is an authentication server that contains authentication information and is trusted to authenticate principals correctly. If SSL is used, then there would be a trusted certification authority that signs cryptographic identity certificates.

Figure 3.8 illustrates how the main authentication components interact.

3.3.5 Security Context Establishment

During the authentication process, the principal has been authenticated and its credentials object, which contains a number of privileges, has been generated. When the principal tries to invoke a remote object, the CORBA security system needs to associate its credentials with the communications and, in particular, transfer the credentials securely to the remote peer. This is done as part of the security association establishment process.

3.3.5.1 Security Contexts

A *security association* denotes "shared security state information that permits secure communications between two entities" [1]. For each security association, a pair of `SecurityContext` objects (one for the client and one for the target) provides the shared state information that represents a security association, such as the credentials used, the target security name, and the session key.

The main purpose of the security association establishment is the secure transfer of security information (e.g., credentials and keys) between communicating parties, because each side normally needs to know for its security enforcement who the other party is and what it is allowed to do. For example, access control on the target side often depends on the caller's identity; hence the target-side security system needs to have access to this information. Establishing the security context can take several exchanges of messages containing security information (e.g., to handle mutual authentication or negotiate security mechanisms).

Once a context exists, it can be used for many subsequent interactions. Note that there is not always a one-to-one relationship between client-target object pairs and security contexts. Contexts may be shared [(e.g., when a client invokes several target objects that have the same (trusted) identity)], and an object can share several contexts with another object (e.g., if a client uses different privileges for different invocations on the same object). During the lifetime of a security context, applications can check its validity (with the operation `is_valid`), and may be able to refresh it (with the operation `refresh`) if permitted.

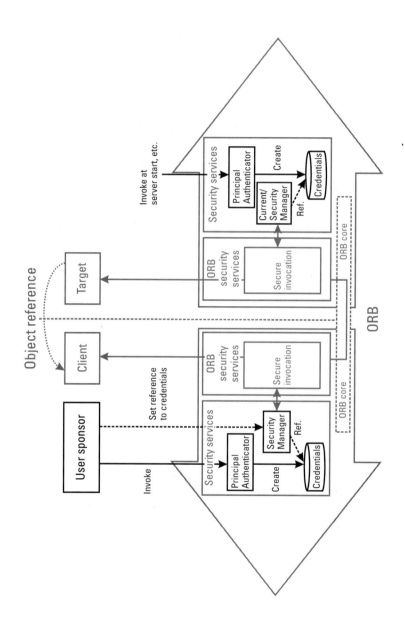

Figure 3.8 Principal authentication.

The security context object stores several different types of credentials:

- The `OwnCredentials` contains the identities and privileges of the local subject associated with the active context.

- The `ReceivedCredentials`, which resides on the target side and contains the identities and privileges of the remote subject from whom the execution context has most recently received a message (if it has received any).

- The `TargetCredential` that resides on the client side and contains a remote principal's authentication information for the client's security context with the server.

- The `InvocationCredentials` contains the identities and privileges the execution context will use the next time it sends a message. This is normally the same as the own credentials, but if the execution context has become a delegate (see Section 3.3.3), then the invocation credential may be the same as the received credential.

3.3.5.2 Context Creation

Most of the actual context establishment work is done by the so-called `Vault` object, which is responsible for creating security context objects and establishing the security association between client and target. The ORB interceptor[5] calls `Vault::init_security_context` to request a security token that is to be sent to the target. The `Vault` then generates the client-side security context object and the token, which essentially contains the `OwnCredential` and some other context information. Once the token is securely transferred to the target side, the ORB security service (interceptor) there calls `Vault::accept_security_context`, which generates the corresponding target-side security context object. The transferred credentials from the token are stored as the `ReceivedCredentials` on the target side.

If the establishment of the security context involves mutual peer authentication or negotiation of security mechanisms, then several token exchanges may be necessary. Note that these exchanges, like the `Vault` object itself and the security context objects it creates, are invisible to all

---

5. ORB services are implemented using interceptors. An interceptor is interposed in the invocation/response path between a client and target object. There are two types of interceptors:
   - Request-level interceptors, which perform transformations on a structured request;
   - Message-level interceptors, which perform transformation on an unstructured buffer.

applications. At the end of such a handshake, both parties have access to a security context object with all related credentials.

In the case of a privilege delegation chain, the `Vault` object associated with the intermediate object can automatically create a new security context when the intermediate object invokes another object. This can trigger another authentication process.

3.3.5.3   Context Access

The ORB architecture provides a standard way to access information associated with the active execution context. An application can find out what execution context it is in and what that context's credentials are by calling the ORB to get its so-called `Current` and `SecurityManager` objects and then querying them to discover any of the credentials associated with the current execution context.

The CORBA security model associates security state information, including the credentials of the active principal, with the `Current` object. So, in essence, the `Current` and `SecurityManager` objects are just another way of accessing the information, such as the credentials for the principals involved, from the active security context.

Figure 3.9 illustrates how the different objects interact.

3.3.5.4   Interoperability

To support the described protocol exchanges necessary to establish a security context (and protect messages, see Section 3.3.6), the CORBA IIOP requires a number of enhancements. To allow for replaceability, these enhancements are added as a separate Secure-Inter-ORB Protocol (SECIOP)[6] that is inserted on top of IIOP and transmits security information and GIOP messages securely across the network (see Figure 3.10). Where possible, SECIOP messages are sent together with IIOP messages rather than as separate exchanges. However, this is not always possible, for example, when a client wishes to authenticate the target before it is prepared to send an IIOP message.

SECIOP uses standardized security tokens to support the establishment of security contexts. Note that, although the type of security tokens is

---

6. The specification also defines a security-enhanced DCE-CIOP protocol that runs on top of DEC's distributed computing environment (DCE). It takes advantage of the integrated security services provided by *DCE authenticated RPC* and provides security features like cryptographically secured mutual authentication, credentials, integrity and confidentiality protection, and protection against replay attacks. It can pass a range of privilege attributes, support controlled delegation, and use secret key technology.

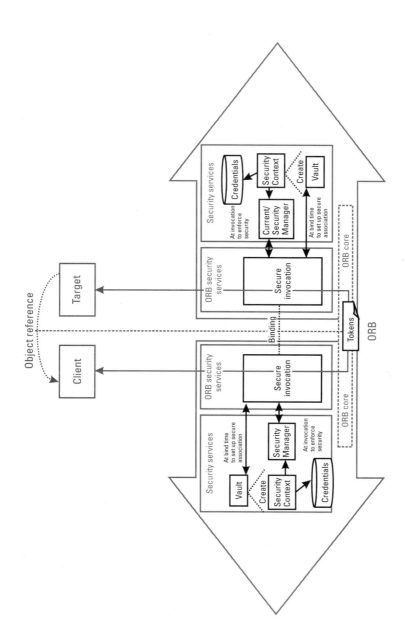

Figure 3.9 Security context establishment.

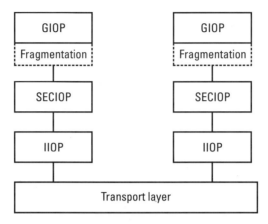

Figure 3.10 SECIOP.

standardized (e.g., establish context, message context), the number of tokens used and their content are mechanism-specific. Token details for the following security mechanisms are specified for use with SECIOP, so that implementations that use the same mechanism (and consistent policies) can interoperate:

- *Kerberos:* This mechanism passes the initiator's identity only for access control and audit and supports delegation with no controls.
- *SESAME:* This mechanism can pass a range of privileges, as well as an identity for access control, and has delegation controls. It has secret and public key technology options and replaceable algorithms, and it uses X.509 V3 certificates with associated certificate management. Only SESAME supports the full CORBA security facilities.
- *SPKM:* This mechanism passes the initiator's identity only and does not support delegation; it uses public key technology. Like SESAME, it uses X.509 V3 certificates and associated certificate management.

Conformant secure interoperable ORBs will support Kerberos and, optionally, SESAME and SPKM. For other mechanisms, other token details may need to be specified to allow for interoperability.

3.3.6   Message Protection

So far, the security system has authenticated the involved parties, created credentials for them, and established a security context between communicating parties (including peer authentication, if required) so that security context information is available to the client and target side security system. At this point, the actual invocation request is sent across the network from the client to the target, and the reply comes back the same way.

Messages can contain valuable information that needs to be protected or, if modified, they can trigger events on the client or target side that lead to a direct or consequential loss. For example, sender or recipient network addresses can be modified to fool either party into committing disastrous actions. Therefore, messages need to be protected against unauthorized disclosure or modification while in transit between a client and target.

3.3.6.1 Policy

Message protection, like many other functional components of the CORBA security model, is enforced based on a policy. The message protection policy specifies what *quality of protection* (QoP) needs to be applied to each message. The security model supports three different kinds of message protection, and the QoP policy defines which of them should be applied to a message, and at what strength:

- *Origin authentication,* which proves the identity of the sender of the message to the recipient. Mechanisms for this type of QoP could be digital signature schemes.

- *Confidentiality,* which prevents unauthorized disclosure of each message data. This is normally done by encrypting the message.

- *Integrity,* which prevents undetected, unauthorized modification of message data and may also detect whether messages are received in the correct order, or if any messages have been added or removed. Examples of mechanism types are cryptographic checksums and sequence numbers.

The security model allows the definition of message protection policies in three places:

- On the ORB layer, system owners can define two kinds of policies to ensure that messages transmitted within their systems are adequately protected: client secure invocation policies, which define the minimum protection that must be applied to messages sent from the system; and target secure invocation policies, which define both the minimum protection that must be applied to received messages and the maximum protection the system can support. This type of policy is attached to the `Current` object (i.e., the active execution context).

- On the application layer, target object owners can define which message protection needs to be applied, so that data flowing into and out of their objects is adequately protected. A description of that policy is written into the target's object reference, so that callers will know the minimum required protection, as well as the maximum supported protection, of the target object.

- Finally, the model supports policies through which principals can control the minimum level of protection that has to be applied to all messages they send and receive. In addition, they may specify the maximum level of protection that they are able or willing to support. This policy is stored in the principal's `Credentials` object.

3.3.6.2 Negotiation

To implement the appropriate level of message protection, the CORBA security system needs to combine the individual policies to arrive at the QoP that has to be applied to a message. This has been included into the model to support a weak form of negotiation between the communication parties. Figure 3.11 illustrates how QoP negotiation works:

- On the sending side, the *required effective QoP* is the collection of the required QoP specified in the principal's credentials, the object reference, and the `Current` object. If the client-side *maximum supported QoP*, which is specified in the object reference and the `Credentials` object, is at least as strong as the required effective QoP, then the security service will create a `SecurityContext` object that implements the required effective QoP. In addition, a context setup token, which contains the required effective QoP, is generated and sent to the target side.

- On the receiving side, the *accepted effective QoP* is the collection of the required QoP specified in `Current` and the credentials object, as well as the sender's applied required effective QoP. If the maximum supported QoP, which is specified in the object reference and the `Credentials` object, is at least as strong as the accepted effective QoP, then the security service will also create a `Security-Context` object that implements the required effective QoP. Otherwise, the security system will refuse to accept the message.

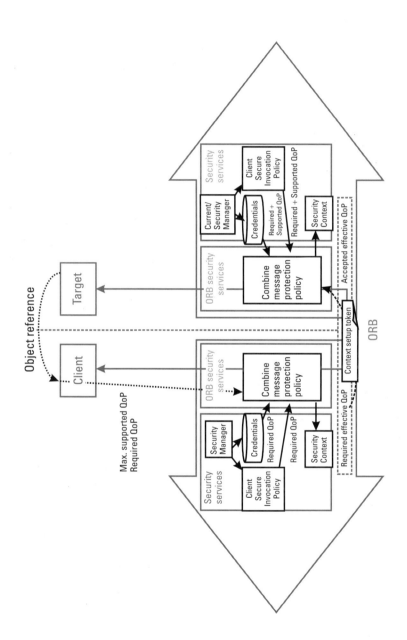

Figure 3.11 Negotiation of message protection policy.

3.3.6.3 Enforcement

Enforcement of message protection is done for each message by simply calling the `SecurityContext` object, which implements the operations `protect_message` and `reclaim_message` for protecting messages for integrity and confidentiality (see Figure 3.12). The security system automatically intercepts all messages and puts them through these functions, so the message protection cannot easily be bypassed.

Message protection is usually implemented using encryption. Since the CORBA security model has to fit a variety of purposes and levels of protection, it allows a choice of cryptographic algorithms for message protection. Furthermore, request and response may be protected differently, and both integrity and confidentiality protection can be applied to the same part of the message. The higher the quality of protection has to be, the better the algorithm needs to be. The effective quality of the protection depends on the quality of the underlying mechanisms. The ORB just makes sure that the encryption cannot be bypassed if it is required for the security association. Note that the underlying cryptography mechanisms are not visible outside the security service—encryption and integrity protection are totally transparent to the user.

Also note that encryption is not necessary if systems provide message protection inherent to the environment (i.e., in the same security environment domain—see Section 3.3.2). For example, virtual private networks (VPN) already provide message protection. The ORB does not provide its own message protection when it operates on such a secure transport layer.

3.3.6.4 Interoperability

To support message protection functionality, the IIOP requires a number of enhancements. The SECIOP (see Section 3.3.5) securely encapsulates the GIOP messages to prevent any modification (e.g., deletion, reordering, replay). Alternatively, the SSL-Inter-ORB Protocol (SSLIOP) can be used to transfer GIOP messages securely on top of SSL, a widely used secure transport mechanism (see Section 3.4.1).

In addition, the IOR format needs to be enhanced to include the authenticated identity of the target, relevant target-side security policy attributes, and the list of security mechanisms supported and required by the target (for QoP negotiation).

Note that secure interoperability is only supported if ORBs share a common interoperability protocol, consistent security policies are enforced, and the same security mechanisms are in use on both sides. In particular, the

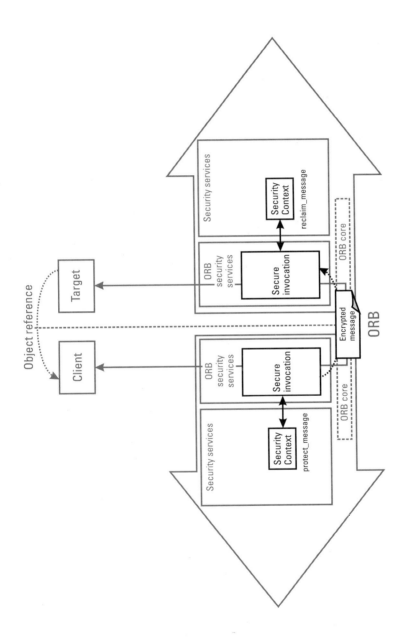

Figure 3.12　Protected invocation.

specification does not define any gateways to translate between underlying security mechanisms (e.g., different encryption algorithms) and security policy domains. In most cases, such gateways would not fit the underlying trust model because they need to have access to the cryptographic keys of all participants to decrypt and reencrypt all traffic.

3.3.7 Access Control

In the previous steps, the security architecture has set the scene for access control. CORBA security has authenticated the involved principals and generated credentials for them. These credentials are then used during security context establishment to set up an authenticated and protected network link between communicating parties, so that the client can securely send an invocation request to the target. On its way, the request can go through a chain of intermediate delegates before it reaches its final destination.

Access control is about restricting access to resources to prevent their unauthorized use [1]. Whenever the request arrives at the target side, the security system there needs to decide if the caller is authorized to invoke the target method. The security model also supports access control at the client side, which can be used to control which requests client objects can send.

Client-side access control is done independently from target-side access control. The client-side ORB enforces the client domain access policy (which checks if the client is allowed to invoke that operation under the given circumstances) using the information from the target object reference, whereas the target-side ORB enforces the target domain access policy (which also checks if that client may invoke that operation under the given circumstances).

Client-side access control is rarely useful in practice. Its purpose is mainly to prevent the network from getting unnecessarily flooded. Access control in the sense previously defined is primarily a target-side concern.

3.3.7.1 Policy

The access control policy describes which target methods each principal is allowed to invoke. To make larger systems manageable, target objects are grouped in security policy domains to which a common policy applies, and principals are clustered into roles, groups, or clearances (see Section 3.2.4). This way, the number of access rules can be reduced dramatically.

But for most access control policies, per-object granularity is not enough. Objects will frequently export a number of operations with differing security needs. For example, an electronic banking object `Bank` could have a method `get_admin_contact` that is accessible to any system user, whereas a `withdraw` operation should be restricted to authorized customers. To write a policy on such a per-operation granularity would not be scalable, as the administer would need to write a separate rule for each operation on each object.

To solve the scalability problems of per-operation granularity, the model associates standard sensitivity levels to each operation. This way, the policy can compare the level required to access the operation with the level granted to the client and only allow access if the client's granted level is sufficient.

The specification provides a standard set of access rights to ensure that both sides interpret access control attributes in the same way. The *rights family* that contains the standard rights includes g (get), s (set), u (use), and m (manage). Several access rights can be combined in two ways: If the "all" combinator applies, then the caller has to possess all rights in the list; for the "any" combinator, any one of the listed rights is sufficient to access the operation. In the `Bank` example, the `get_admin_contact` operation could be accessible with *any* (g,s,u,m), whereas the `Withdraw` operation could only be invoked if the caller has *all* (g,s,u).

In summary, the standard access decision model bases its access decision on the following:

- The current privileges of the caller (e.g., access ID, roles, groups, security clearance);

- Any control that is applied to privileges (e.g., lifetime);

- The operation to be invoked;

- The access policies of the target object.

Implementations may define additional rights families to fit the model to their particular access control requirements. This way, the access control model can support a number of different policies, such as access control lists (ACL), capability lists, and role-based access control (RBAC). However, the use of additional rights families may impede interoperability, since the target side may not be able to interpret the caller's privileges correctly.

3.3.7.2 Evaluation

The access control evaluation functionality is encapsulated within the `AccessDecision` object:

- Whenever a message arrives, the ORB security service intercepts it and passes it to the access decision object to find out if the access is allowed.

- The access decision object then forwards the caller's credentials (from the `Current` object) to `DomainAccessPolicy`, which returns the granted rights for the calling principal.

- `AccessDecision` then calls the `RequiredRights` object to find out the required rights for invoking the target method on the target object.

- Access decision is now able to compare the granted rights with the required rights. It will only allow the invocation if the granted rights match the required rights; otherwise the request will be blocked. Additional checks can be put in place at this point, such as controls that are applied to the privileges (e.g., lifetime).

Figure 3.13 illustrates the access policy evaluation process.

3.3.7.3   ORB Layer and Application Layer Access Control

So far, we have only described how access control works on the ORB layer. At that layer, access control can be applied to both security-aware and security-unaware applications. In addition, the access controls are built into the ORB message path and therefore cannot be easily bypassed.

In addition to ORB layer access control, the CORBA security architecture allows security-aware applications to evaluate and enforce their own application-specific access control policies. Both the client and target application can call the `DomainManager` object of the underlying CORBA security system, which uses the policy information in `Policy` objects to decide if the invocation should be granted or not. The domain manager gets the caller's privileges from the `Current` context object. Figure 3.14 illustrates the difference between ORB layer and application layer access control. Both client and server side show ORB layer access control. The client side also shows how application layer access control is done if policies are managed in domains. The `AccessDecision` object gets the relevant references to the access control policy from the `Current` object. The target side illustrates access control for applications that manage their own policies.

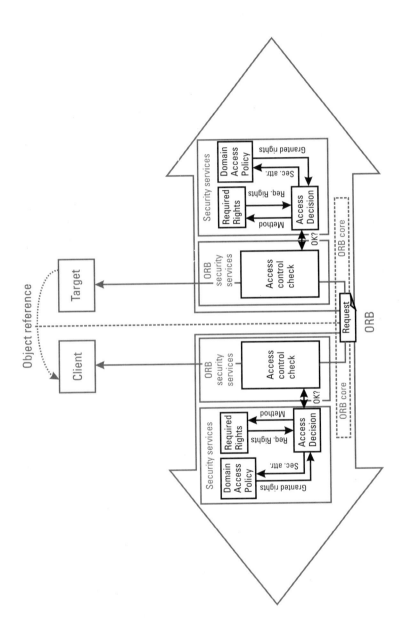

Figure 3.13 Access control evaluation.

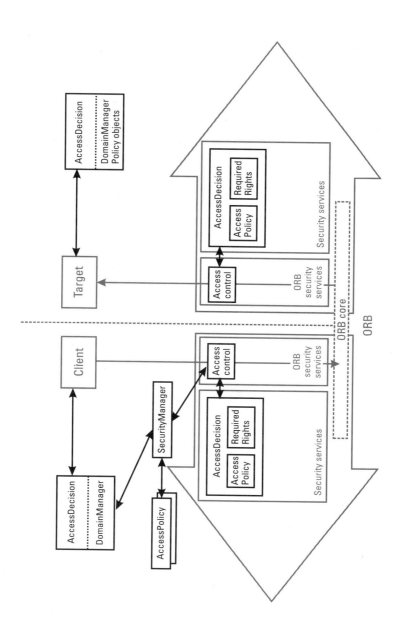

Figure 3.14 Application and ORB layer access control.

3.3.7.4 Enforcement

The `AccessDecision` object does not directly rely on any underlying security mechanisms for its evaluation, since it uses standardized access rights. The enforcement is automatic, as long as the security service intercepts all messages at the message path and mediates them through `AccessDecision`.

The implementations of `DomainAccessPolicy` and `Required-Rights`, on the other hand, select access rights based on (mechanism-specific) credentials provided by the authentication, which were transferred during security context establishment and safeguarded by the message protection functions previously described. For example, if the access decision depends on the caller's identity, then the authentication function needs to be trusted to generate the correct credentials. In addition, the security context establishment needs to be assumed to transfer the client credentials to the target side in a trustworthy manner (i.e., without being tampered with). Therefore, the access control will only be as strong as the authentication, security context establishment, and message protection functions. Note that this conclusion is not specific to CORBA security; it also applies to most access control systems.

Also, as with all the functional components described so far, the CORBA access model assumes a trusted ORB. Every application object must trust its underlying ORB, as well as all other underlying system components (e.g., the operating system) to work as it should.

3.3.8 Security Audit

Security audit is the process of recording details about security-relevant events in an audit log. Audit logs store event details in such a way that they cannot be modified or tampered with. An audit event is the data collected about an event in the system for inclusion in the system audit log or in an alarm notification. For example, the data that should be recorded for particular method invocation could include the called operation, the caller identities and privileges, and the time of the invocation.

By analyzing these security audit logs, auditors can detect actual or attempted security violations. They can also see from the logs whether the protection measures are adequate and if they comply with established security policy and operational procedures.

3.3.8.1 Policy

If everything that happens in the system was recorded (regardless of whether it is relevant or not), then the audit log would become very large and filled up with irrelevant audit events. Consequently, the system performance would deteriorate, and later analysis of the audit logs would be difficult. Hence, audit policies are used to restrict what types of event logs are to be generated under which circumstances.

The CORBA security system stores audit policies in `AuditPolicy` objects, which are called by the `AuditDecision` object to see whether an event has to be generated. The audit model defines a number of *audit selectors* for particular system events. Each time an event occurs, the audit service looks for the corresponding selector in the audit policy and generates an event only if it finds the corresponding selector; otherwise, no event will be created. The specification currently supports selectors for the following security event types:

- When principal authentication takes place (e.g., when users log on);
- When secure connections are established;
- When access control checks are enforced;
- When an invocation occurs;
- When changes in the security environment occur;
- When changes in the security policy occur;
- When objects are created;
- When objects are destroyed;
- When nonrepudiation evidence is generated;
- When all events are to be recorded.

The model allows the definition of additional customized event type selectors to fit to more application-specific requirements (e.g., additionally defined security attributes like the peer's network address).

Each selector type in the policy contains a selector value list, which states the circumstances under which the event will be generated. The value list will specify one or more (combined with "all" or "any," as discussed in Section 3.3.7) of the following values:

- The target object interface type;
- The target object reference;
- The invoked method;
- The initiator's credentials;

- Whether the operation succeeded or failed[7];

- The time of the event;

- The day of the week.

In summary, the audit policy describes the conditions that trigger the audit service to generate a log entry. The description is based on a number of event categories that contain a number of values. These values restrict the event generation to the relevant events.

3.3.8.2   ORB Layer and Application Layer Audit

Audit policies can be defined for both the ORB layer and application layer. ORB layer audit policies (client invocation audit policy, target invocation audit policy) control, on the client and target side, what events are recorded as the result of relevant system activities. ORB layer audit policies are enforced transparently by the security system—this way, security-relevant audit logs can be generated for security-unaware applications.

Application audit policies control which events are to be audited by security-aware applications. Which types of events should be logged depends on the particular application; therefore, it is often necessary to introduce additional customized event types. For example, one could have an event type "money transfer," which creates a log every time money is transferred into or out of the application.

Figure 3.15 illustrates the main components involved in auditing. The client application side shows how auditing is done if audit policies are managed using domains. The `AuditDecision` object gets the relevant references to the domain audit policies from the `Current` object. The target application side shows auditing for applications that manage their own audit policies.

3.3.8.3   Policy Evaluation

Each time a request is sent, the client-side security service intercepts it and calls the `AuditDecision` object to find out whether an audit event should be generated. The `AuditDecision` object then compares the selectors in the client invocation audit policy for the current execution context with the request and event data (e.g., time). If the request attributes match the selectors and values, then audit decision returns a positive answer.

---

7. It is not clear what this means, in particular for ORB layer audit.

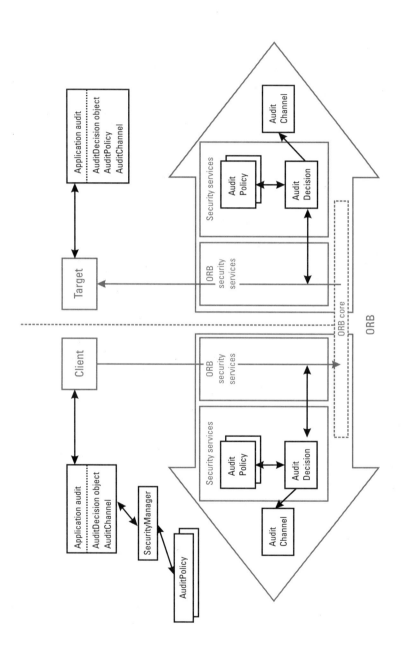

Figure 3.15 ORB layer and application layer security audit.

When the request arrives at the target side, the security service there also intercepts it and calls its `AuditDecision` object. On the target side, the request and event data are compared to the relevant audit selectors in the target invocation audit policy for the target object.

Application objects have to call `AuditDecision` themselves whenever they send or receive a message, and the audit decision is based on their application audit policy.

3.3.8.4 Enforcement: Audit Channels

If the `AuditDecision` object returns a positive answer, then the ORB or application needs to generate an audit event. This is done by passing the event data to an associated `AuditChannel` object, which then stores the logs in an implementation-specific way. To locate the correct audit channel, the ORB or application can query their `AuditDecision` object, which contains a reference to the audit channel object associated with the policy.

The event data that is passed to the `AuditChannel` object includes:

- The event type;

- The event initiator's credentials;

- The time the event occurred;

- The selector values used during the audit decision;

- Other event-specific data (optional).

The audit model only specifies how to pass event data to the `AuditChannel`. Within the `AuditChannel`, this event data can be used in many implementation-specific ways. Events can be either recorded on disk for later analysis or alarms can be sent to an administrator.

Because of the implementation-specific nature of the `AuditChannel`, the specification does not state how audit logs are administered (i.e., how audit records are filtered later, how audit trails and channels are kept secure, and how records can be collected and analyzed).

3.3.9 Nonrepudiation

One of the most complex requirements of the CORBA security architecture is to provide the means to make principals accountable for their actions. The security audit logs already described (see Section 3.3.8) are only part of the solution, because each node can only create audit logs for its own side of the communications, and no one can prove to the other side (or to a trusted arbitrator) that the local logs have not been modified with malicious intent. For example, the target can delete audit events for a particular invocation from the audit logs on the target side if a caller tries to prove that it initiated an invocation that was never carried out. Even worse, if a caller gets billed for a service associated with a particular invocation, then it cannot prove that it did not invoke the service if a false entry was inserted into the audit log.

3.3.9.1 Evidence

To solve the problem of disputes, the security model needs to provide functionality to create irrefutable evidence that prevents participants in an action from convincingly denying their responsibility for the action or from making principals accountable for actions in which they were not involved. This functionality, which is called nonrepudiation, is based on digitally-signed *evidence tokens* that are generated by the nonrepudiation service to prove that a particular principal initiated a specific action. The nonrepudiation service should give all participants the means to cast doubt on all false accusations that are made against it (because no valid evidence is available on the accuser's side), and to support all true accusations it wants to make (because valid evidence is available on the accuser's side).

The CORBA nonrepudiation model distinguishes normal evidence tokens from *complete evidence tokens*, which—in addition to the signature—contain all information relevant in a future dispute but which may be unavailable by the time the verification takes place. In the simplest case, this can include the signer's digital certificate to prove that the key was valid at the time of the evidence generation. In more complex cases in which the environmental parameters change dynamically, the complete evidence token could also contain the actual message sent, as well as some information about the state of the system (e.g., the time).

The model supports evidence generation for a number of different kinds of actions. The most commonly used evidence token types are:

- *Proof of origin*, which identifies the originator of a message;
- *Proof of receipt*, which establishes that a message has been received by a particular party.

3.3.9.2 Policy

The client and target nonrepudiation policies specify the types of evidence that should be generated for each message. Evidence generation cannot be provided transparently on the ORB layer because nonrepudiation has to cast liability on principals and not on the underlying ORB. To ensure account-ability of the principal (or legal entity, such as a service provider) behind a used application, the ORB must not have access to the principal's evidence signing key. Otherwise, the principal could claim that the underlying CORBA security system abused his key to create false evidence. This measure is necessary because the digital signatures used to sign the evidence tokens are generally only binding if they were applied with the explicit consent of the signer, and if only the owner of the key (i.e., the principal who will be held liable) could have signed the message.

3.3.9.3 Evidence Generation and Verification

Application objects can use the `NRCredentials` object to generate and ver-ify evidence tokens, get token details, and set or retrieve policy details. Gen-erally, digital signature schemes can generate evidence and protect its integrity, but the CORBA security model does not mandate any particular technology to be used in the `NRCredentials` implementation. Figure 3.16 illustrates where the proof of origin and proof of receipt are generated and verified for a message transmitted from client to target.

3.3.9.4 Dispute

When a dispute occurs, an independent third party (often called the adjudi-cator) needs to settle the disagreement based on the nonrepudiation evidence provided by both parties and in accordance with its policy. The policy has to

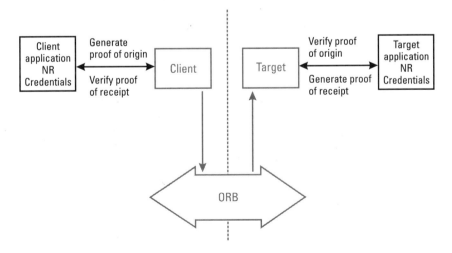

Figure 3.16 Evidence generation and verification.

be accepted by all participants and can have a legal basis, such as laws or con-
tractual terms.

The security model does not define how accusations are carried out
or how disputes are settled. One possible implementation could include an
adjudicator server somewhere on the network that can be called by any party
to accuse other parties. This invocation could contain all relevant evidence
tokens and a reference to the accused party. The adjudicator could then
check the validity of the evidence and—if the evidence (e.g., the cryptographic keys and certificates) is valid—contact the accused party, ask for their
evidence, and present both parties with the verdict.

3.4 Secure CORBA on the Internet

CORBA was originally developed for Intranet-centric environments where
legacy applications (e.g., back-end data stores) had to be integrated with
state-of-the-art technology (e.g., an enterprisewide controlling application).
In these early days, CORBA was successfully implemented within rather
static environments with well-defined ownership, simple trust relationships,
and central management (e.g., for large manufacturing systems). Security in
such systems was well understood and rather easy to implement and
administer.

With the unexpected growth of the Internet in the mid-1990s, more
and more applications were designed to be used by a large customer base.
CORBA seemed to be an ideal technology to handle these very large,
dynamic, and distributed applications on very heterogeneous platforms.
When electronic commerce applications became increasingly commonplace,
security was a major concern. It soon turned out that the Intranet-centric
CORBA security model (which was first published in 1995) was not applicable in many cases, partly because it was too heavy to be downloaded with the
client application in a Java applet, and partly because the system could not
cater to the new security requirements. For example, now there was mutual
suspicion between the participants that operated on a potentially hostile
underlying platform, and central security administration was not possible
anymore in these highly dispersed systems. In addition, several (untrusted)
firewalls were in the communications path between the client and target,
which broke CORBA's end-to-end security model. There is an attempt to
support firewalls within the CORBA security model, which will be described
in this section.

To meet the growing demand for Internet security, the industry integrated the lightweight SSL security protocol into their ORBs in a proprietary

fashion. This protocol had been in use for many years to secure Internet traffic. The OMG reacted to the growing number of incompatible SSL-based products by standardizing the SSL-based SSL-Inter-ORB-Protocol (SSLIOP), which can be used instead of SECIOP.

Toward the end of the 1990s, an increasing number of Internet users started using mobile wireless devices such as palm-size organizers and mobile phones to access services on the Internet. Due to resource restrictions, the development of demanding applications on such devices is still a challenge. Another problem is integrating such highly heterogeneous components into the rest of the system. Today, telecommunications providers and pocket computer manufacturers acknowledge the need for a common secure application platform, and CORBA could be the architecture of choice. But this depends on the ability of the OMG and its members to redesign the specifications (in particular, the security architecture) to meet these new requirements. Our MICOSec implementation, which is presented in Chapters 4 through 7, was modified to fit those needs and can be used on an off-the-shelf pocket computer.

3.4.1 SSL/TLS

SSL/TLS differs from most other security mechanisms used with CORBA. While mechanisms like Kerberos or SESAME provide a number of security functions through an interface (e.g., GSS-API) that can be integrated into CORBA's SECIOP, the SSL protocol is a secure transport mechanism that uses network sockets directly as its interface. In other words, SSL can be set up by the layers above and then automatically enforces security on all traffic that goes through a particular socket.

To solve the problem that SSL cannot be integrated into the CORBA security architecture like other security mechanisms, the CORBA community came up with the idea of *pluggable protocols*. The new SSLIOP was introduced as an alternative transport mechanism for IIOP (i.e., replaces the transport layer of the CORBA architecture). This way, the CORBA security architecture can set up the SSL connection and then use SSLIOP to communicate on top of it.

MICO, the ORB used as the basis for MICOSec, provides a proprietary SSL extension, which is reused within MICOSec. Note that most ORB products support SSL, but because of the late standardization of SSLIOP, products from different vendors are sometimes incompatible when SSL is used.

SSL uses standard X.509 cryptographic certificates as identity tokens, which need to be administered. Due to the widespread use of SSL to secure

Web traffic, a number of off-the-shelf PKI products are available to generate, verify, update, and revoke SSL keys and certificates.

3.4.2 Firewalls

The CORBA security architecture assumes that there is a direct network connection between client and server over which protected IIOP requests and replies can be sent. All security policies are enforced end-to-end by the ORB layer security services (or by security-aware applications) on the client and target side.

Unfortunately, the situation is quite different for many real-world environments, where clients and targets reside in subnetworks with different security policies in such environments. Firewalls enforce a security policy on all network traffic that crosses the domain boundaries [18]. Objects within the domain protected by a firewall are said to be within the same enclave. A firewall controls which callers from outside are allowed to communicate with which targets within an enclave, and which protocols and applications are permitted for communication. In addition, it needs to prevent attacks on the objects within the enclave.

Figure 3.17 shows a typical case where two objects in segregated enterprise enclaves communicate over the Internet. Both enclaves are each protected by firewalls that have to be traversed by all traffic between the client and target enclave. When a client wants to invoke a target, it needs to contact its firewall, which enforces a security policy and then contacts the target

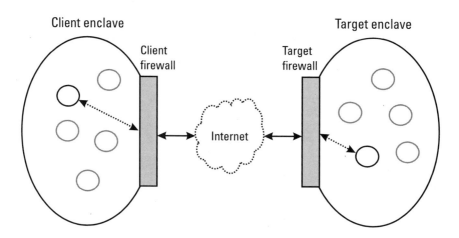

Figure 3.17 Invocation across firewalls.

firewall. The target firewall also enforces its security policy and, provided the access is granted, relays the traffic to the target.

3.4.2.1 Problem Definition

The CORBA secure interoperability model and its protocols do not easily integrate with firewalls for a number of reasons: First, there are a number of conflicting interests. CORBA's security model is based on end-to-end security where components defend themselves (hence, often referred to as "the self-defense model of object-oriented computing"), while the firewall approach controls access to relatively unprotected objects at the enclave boundary ("hard shell and soft inside model"). In the firewall security model, all network traffic is unprotected and individual objects do not enforce their own security policy. Instead, the firewall controls access to the objects within its enclave, based on the caller identifier, the message content, and the target to be invoked. In CORBA's end-to-end security model, the client establishes a protected security context with the target that allows peers to authenticate themselves and exchange protected messages. Firewalls on the communications path cannot validate and filter the traffic because the protected messages are not accessible as plaintext (see Figure 3.18).

Apart from this fundamental clash, there are also several difficulties of a more technical nature. First, objects are addressed by their IOR, which contains the host name and port number of the server. If firewalls are in the communications path, then in some cases the IOR needs to point to the firewall and contain some information that tells the firewall which target to contact.

In addition, CORBA allows TCP ports to be dynamically assigned to target servers (e.g., a new port can be randomly chosen for each object generated by a factory object). The client gets this port number from the IOR, but the firewall also needs to know this to enforce its access control policy. In particular, the firewall needs to know when an object terminates to make

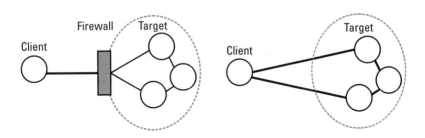

Figure 3.18 Firewall security versus end-to-end security.

sure that calls to its corresponding port are rejected. This is particularly important if a new object with different security requirements reuses that particular port.

Another problem is that CORBA's callback features blur the distinction between client and target, as targets can call back clients at some point, for example, for asynchronous notifications, and client-side firewalls are normally configured to reject incoming requests to protect clients from network attacks.

Moreover, traffic can go through several enclaves with differing firewalls, for example, if different departments within an enterprise have their own firewalls to protect their subnetwork.

3.4.2.2 Toward a Solution

On the one extreme, a widespread and technically simple solution involves tunneling CORBA traffic through well-known ports over well-known protocols such as the Hyper Text Transfer Protocol (HTTP) or its secure version HTTPS. Its advantage is that it is largely transparent and does not require any major modifications to the ORB or security architecture, but it has many downsides. For starters, it cannot solve any of the technical difficulties previously described. But more importantly, it bypasses the enforcement of the firewall security policy and therefore jeopardizes the whole enclave.

The other extreme involves compromising end-to-end security by forcing clients to establish a CORBA security context with the firewall instead of the target. The firewall then establishes another security context with the target if the access should be granted (such a firewall is called *IIOP layer proxy*). However, this solution has a number of drawbacks. First, it breaks end-to-end authentication and message protection, as all parties have to trust all firewalls on the message path to keep the keys and message data secret. But there are also functional weaknesses, as the firewall can only interpret the IIOP header and the message header, whereas the potentially dangerous message content is only an unstructured byte stream (it can only be understood by the corresponding stubs and skeletons) and, therefore, cannot be validated.

A trade-off between these radical solutions is to preserve end-to-end security and transparently provide some basic filtering of the underlying TCP stream (such a firewall is called *TCP layer proxy*). Making sure that callers from certain network addresses can only access a particular port on a particular host protects the rest of the enclave from attacks (which often is a more important requirement than protecting the actual CORBA server). The target object then defends itself from IIOP layer attacks (with the CORBA security services), while the TCP layer proxy prevents many classical TCP layer attacks by rewriting all packets. On the client side, the widely supported Socket Server (SOCKS) library can be integrated into the client network stack so that the client traffic is transparently sent to the SOCKS server, which passes it on to the target. With SOCKS, the client can dynamically choose the target of the connection. On the server side, different TCP layer proxies are preferable, as static port mapping is normally required.

Note that these approaches require a number of changes to the ORB, the IOR, and the CORBA protocols to solve (some of) the technical difficulties described.

3.5 Conformance

Having a sound security architecture is a first and important step toward a secure CORBA system, but there are many other factors that also determine whether the resulting system is secure or not. Of particular importance is the quality of the CORBA security services implementation, as well as its integration with underlying security mechanisms.

It soon became clear that most CORBA security vendors were not able or willing to implement products that cover the whole range of specified security features, partly because the development was deemed too difficult and expensive, and partly because the resulting products were considered unsuitable for most customers' needs.

Consequently, it was decided early on by the OMG that not all products need to support all the facilities defined in the CORBA security specification. Instead, two levels of conformance were specified (see Figure 3.19). For the first level of conformance (level 1 security), it is mandatory to provide a specified set of security features during object invocation for security-unaware applications and to allow security-aware applications to enforce their own access control and auditing. The second conformance level (level 2 security) includes a wider range of security facilities and allows applications

Main functionality level		Functional option	Security replaceability				Security interoperability	
1	2	Non-repudiation	ORB services	Security services	Security ready ORB services	Security ready security services	SECIOP, CSI 1, Kerberos + MD5	SECIOP, DCE-CIOP, Kerberos

Figure 3.19 Conformance levels and options.

to control the security provided at object invocation. It also includes security policy administration. Further conformance options specify nonrepudiation, security service replaceability, and secure interoperability.

It is important for customers to know the level to which a CORBA security product conforms and which options are supported, because buying a CORBA security product does not automatically mean that the whole specification has been implemented, and it does not at all imply that the product is secure. There have been cases in the past where nonconformant security products were advertised as CORBA security products, even though the supported functionality differed significantly from the specification. This section summarizes the different conformance details to help you choose the right product for your particular needs.

Note that these conformance levels only specify *what* has to be implemented (i.e., which interfaces). They do not detail *how* the different parts of the model should be implemented to be effective.

3.5.1   Level 1 Security

Level 1 conformant products provide all applications (i.e., both security-unaware and security-aware) with at least the following functionality:

- Principal authentication inside *or* outside the object system;
- Secure invocation between client and target object (including unilateral authentication, integrity, and/or confidentiality) on the ORB layer *or* outside the object system;
- Simple delegation of client security attributes to targets;
- ORB-enforced access control checks, with support for domains and roles but no support for administration;
- Auditing of security-relevant system events (but not by object invocation).

In addition, security-aware applications can use the level 1 interfaces to retrieve security attributes, which they may use to enforce their own security policies (e.g., access control based on the application state). The level 1 interfaces allow access to security options and attribute details (`get_service_information` and `get_attributes` on `Current`). No administrative interfaces are mandatory at this level.

This conformance level specifies the minimum requirements for level 1 conformance. Of course, level 1 conformant implementations can optionally provide any of the other security features in the specification.

3.5.2   Level 2 Security

For level 2 conformance, security services need to support extra functionality on top of the level 1 functionality. For both security-unaware and security-aware applications, conformant products need to support:

- Principal authentication both inside *and* outside the object system;

- Additional secure invocation features, in particular, peer authentication and message protection at the ORB level. Further integrity options such as replay/reorder protection can be requested (but need not be supported). Also, the standard `DomainAccessPolicy` (for access control) and selective auditing have to work on a per-operation granularity.

At level 2, security-aware applications can control in more detail:

- *Options for secure invocation:* Applications must be able to choose the quality of protection of messages required, change the privileges in credentials, and choose which credentials are to be used for object invocations. They should also be able to specify whether credentials are to be used only at the target (e.g., for access control) or whether they can also be delegated.

- *Delegation:* The application can request (unspecific) "composite" delegation, and the target can obtain all credentials passed, provided all participating nodes support this.

For policy administration, all security policy types except nonrepudiation have to be supported, and the standard policy management interfaces for each of the level 2 policies have to be implemented. ORBs and applications must be able to find out what ORB layer security policies apply to them. Applications must also be able to locate and use their application layer policies to make decisions about what security is needed or to get the information needed to enforce the policy.

Level 2 conformant applications have to implement all application interfaces (except for nonrepudiation, which is optional), all security policy administration interfaces, and all administrator's interfaces. However, this does not mean that all specified values of privilege attributes, delegation modes, and communications options have to be implemented. Instead, some of these interfaces may raise a `CORBA::NO-IMPLEMENT` exception.

As with level 1, CORBA security services implementations that conform to level 2 can optionally provide any other specified security features.

3.5.3   Nonrepudiation Option

An ORB may also conform to the optional security functionality defined in the specification. Currently, only nonrepudiation is specified. To conform to the nonrepudiation option, all interfaces for evidence generation and verification (`NRCredentials`) and the nonrepudiation policy (`NRPolicy`) have to be implemented.

Note that it is not necessary to fully conform to level 2 in order to be able to support the nonrepudiation option.

3.5.4   Security Replaceability

Security replaceability specifies the requirements to support the integration of ORBs and security services from different vendors. If an implementation does not conform to security replaceability, then the specified standard security policies or security services implementation cannot be replaced. There are two aspects to security replaceability:[8]

- *Security features replaceability* allows the security features to be replaced. This requires an ORB (or the ORB Services it uses) to use the implementation-level security interfaces defined in the specification, such as the `Vault`, `SecurityContext`, `AccessDecision`, `Audit`, and `PrincipalAuthenticator` objects.

- *ORB services replaceability* is about segregating the ORB from the security services (and ORB services in general) as much as possible, so that a variety of differing ORB services products can be integrated with the ORB. This involves using standardized interceptors to integrate the security services into the ORB, and using the standard operation `get_policy` (and its associated security policy operations).

---

8. There is also a weaker third option: An ORB is said to be *security ready for replaceability* if it does not provide any security functionality itself, but does support one of the security replaceability options. This means that later integration with conformant security services is supported.

3.5.5 CSI

First, conformant ORBs must be able to generate and use the security-enhanced IORs with the specified security tags. Second, conformant ORBs must use one of the specified security protocols (e.g., the SECIOP or SSLIOP) to transmit and receive the security tokens needed to establish security associations and the protected messages used for protected messages once the association has been established.

CSI is specified in three levels of incremental functionality, and implementers can choose to support the option most appropriate for their product:

- *CSI level 0* includes identity-based policies without delegation; access and audit policies are based on the identity of the initiator.

- *CSI level 1* contains identity-based policies with unrestricted delegation (i.e., impersonation). Access and audit policies are based on the identity of the initiator or delegate (depending on the delegation policy).

- *CSI level 2* involves identity- and privilege-based policies with controlled delegation. A wider range of access and audit policies are supported (e.g., role-based access controls and mandatory access controls).

ORBs can only interoperate securely if they use the same security mechanisms (or use a bridge between them, if available) and specify all the cryptographic profiles they support. Therefore, for each CSI level, a set of standard security mechanisms and algorithms has to be supported with SECIOP: SPKM for CSI level 0, KerberosV5 for CSI level 0 or 1, and CSI-ECMA (public key, secret key, or hybrid) for CSI level 0, 1, or 2. Alternatively, SSL (implementing the SSLIOP) or DCE (implementing both the DCE-CIOP and SECIOP protocols) can be used as underlying security mechanisms in conformant products. Figure 3.20 summarizes which mechanisms can be used at each CSI level.

To overcome weaknesses in the underlying secure communications mechanisms (e.g., SSL's missing support for delegation), the OMG specified an additional Security Attribute Service (SAS) as part of the *CSIv2* architecture [17]. It introduces an additional security protocol layer on top of the underlying secure transport mechanism that provides client authentication, delegation, and privilege token functionality. The SAS protocol is modeled

	IIOP				DCE-CIOP
	SECIOP			SSLIOP	DCE security
	SPKM	Kerberos5	CSI-ECMA	SSL	
CSI 0	✓	✓	✓	✓	✓
CSI 1	✗	✓	✓	✗	✓
CSI 2	✗	✗	✓	✗	✓

Figure 3.20 CSI levels.

after the Generic Security Service API (GSSAPI) token exchange paradigm [19] and exchanges its protocol elements in the GIOP service context.

In essence, the SAS protocol allows tokens to be exchanged across a secure underlying transport. The X.509 identity tokens exchanged at the attribute layer allow an intermediate to act on behalf of (i.e., impersonate) some identity other than its own. To accept such a delegated identity, the target either has to trust the intermediate directly or base its trust on a proxy rule certificate (called authorization token [20]) that has been signed either by the initiator or a trusted privilege authority. Such a proxy certificate specifies whether or not the intermediate is authorized to act on behalf of the initiator. More details on the CSIv2-SAS delegation protocol can be found in Section 6.7.

Although CSIv2 is security related, it is not specified within the CORBA security services specification, but as part of CORBA v2.4. At the time of this writing, the relatively new CSIv2 protocol is hardly used in practice, but it is likely that it will be widely used in the near future; first, because it runs on top of the extremely widely-used SSL protocol, and secondly because it is supported by both CORBA and Enterprise Java Beans (EJB) and, thus, enables secure interoperability between both technologies.

3.6 Features or Wish List?

Now that the main design goals (see Section 3.2) have been introduced, as well as a conceptual overview of the security architecture (see Sections 3.3 and 3.4), it is time to take a brief reality check. Does the security architecture live up to its design goals? Can it live up to all of them?

The list of design goals is very ambitious and has to be understood as a wish list—if all these goals could be achieved, then most of the problems of distributed systems research and information security research would be solved. But the mission of the OMG and its members is not to get deeply involved in fundamental research but to produce specifications for a standard middleware architecture.

One of the main difficulties is that some goals conflict with others. We already mentioned that there is a clash between interoperability and flexibility [2], because flexibility involves the customization of functionality, whereas interoperability requires some standardized functionality. Also, effective assurance can only be achieved (and certified) by looking at the system as a whole, whereas portability means that components can be changed without affecting the layers above. It is clear that the security architecture can only try to find the best trade-off between such conflicting goals.

Third, some design goals depend on the actual implementation of the ORB and CORBA security services product, in particular, the assurance, performance, and scalability goals. Note that OMG specifications only specify object interfaces but do not dictate how objects are to be implemented. This way, OMG specifications allow for a wide range of possible implementations but, while being able to prevent major obstacles, cannot ultimately ensure performance and scalability for concrete implementations

How secure the resulting system is also depends heavily on the effectiveness of the underlying security mechanisms. CORBA security is only as strong as the weakest mechanism (i.e., the resulting system is vulnerable if any bugs exist in the security mechanisms, even if the CORBA security architecture has been correctly implemented).

We will now briefly examine potential shortcomings of the CORBA security model for each of the main design goals introduced in Section 3.2.

3.6.1   Interoperability

By specifying common attribute types, access rights families, and audit selector types, the model tries to support consistent security policies across different ORB and security services products. In addition, the CSI specification specifies a common set of security mechanisms, as well as standard security-enhanced protocols and object references. But despite that, security policies across differing security mechanisms are often inconsistent because the security attribute content within policies (or on-the-wire tokens) is mostly security mechanism–specific. The reasoning behind this is that mechanism-specific attributes often cannot be abstracted without semantic mismatches or granularity problems [3].

Interoperability between nodes with differing security attributes or differing underlying security mechanisms require mechanism-specific bridges.

These implementation-specific bridges, which are not part of the specification, need to be secured and trusted to map correctly. They are difficult to design and implement and introduce the same potential semantic and granularity problems previously mentioned.

3.6.2   Transparency and Abstraction

The single sign-on feature can make CORBA security almost invisible for users. From an application programmer's perspective, ORB layer security features can be (almost) transparently enforced for security-unaware applications. Applications do not even need to know the nature of the underlying security mechanisms, because all policies are handled outside the application on the middleware layer.

For administrators, CORBA security is not transparent. This is not a problem, as administrators have to deeply understand the system to be able to specify appropriate policies. However, the fact that CORBA security cannot effectively abstract the content of security attributes from underlying security mechanisms creates administrative problems. Central policy administration tools have to be able to handle all types of (mechanism-specific) attribute content, and attribute content will need to be updated when security mechanisms at a node in the system are changed.

The attempt to provide some level of attribute abstraction by using interface types to describe objects, instead of the particular object instance, makes the situation worse. This is partly because interface names in CORBA cannot be obtained reliably for various reasons (e.g., interface inheritance), and partly because interface types do not describe the identities of object (instances) well enough.

3.6.3   Flexibility, Portability, Integration

We previously described why flexibility clashes with interoperability. To support interoperability, CSI specifies a number of standard mechanisms. These mechanisms have to be supported on all participating nodes and cannot be modified without changing them everywhere and without also modifying the attribute content in the policies. This inhibits mechanism flexibility and code portability.

Flexibility and portability also clash with assurance. Assurance can only be certified by analyzing the system as a whole, and the certification needs to be reevaluated whenever system components are modified. This clashes with portability, which promotes that components can be changed without affecting the rest of the system.

The integration of SSL shows that not all security mechanisms can be integrated into the architecture—the work-around of using SSL as an alternative transport does not integrate with various core components of the model (e.g., the `Vault` object).

3.6.4   Scalability

The upper limit of users, objects, policy entries, etc., depends on the actual implementation. From an administration perspective, scalability is supported in CORBA security through domains and roles (or groups, clearances, etc.), but no tools are specified to manage them. Therefore, it is not possible to manage CORBA systems in an interoperable way, which ultimately inhibits scalability. Note that management of cryptographic keys also needs to be scalable, which depends largely on the particular PKI implementation and on the way it is integrated into CORBA.

The upcoming *Security Domain Membership Management Service* [10] and CORBA PKI [11] specifications should help mitigate this problem.

3.6.5   Reliability and Assurance

Automatic security enforcement by intercepting all traffic on the ORB message path (through interceptors) is a sound architectural feature, but its reliability depends to a large extent on the style and quality of the actual implementation. Real assurance is difficult to certify for applications built on top of the CORBA security model, because a lot of components on different layers of the architecture play together to enforce security, and because there is no small trustworthy security kernel in the CORBA security model.

Also, the flexibility and portability goals, which encourage frequent "plug and play" of different components, jeopardize the system's reliability as a whole. This is because combining components that are individually certified to be secure does not automatically imply that the resulting system is also secure [21]. To solve this problem, it would be necessary to reevaluate and recertify the whole system after every modification.

3.6.6   Simplicity

The CORBA security architecture is not simple and easy to understand for many reasons. First, it tries to cater to many application requirements with a one-fits-all architecture, and, in particular, it tries to trade-off many conflicting design goals. Also, security enforcement in distributed object systems is inherently complex because systems are large, dynamic, and heterogeneous.

Abstracting in such systems is not easy and, thus, the architecture gains further complexity.

Administration of the CORBA security architecture is complicated because of its support for many different policy and mechanism types, and because there is often mutual distrust between participants.

3.7 Summary

In order to be useful, the CORBA security architecture has to fit to CORBA in a nonobtrusive way, which means that it has to preserve the main CORBA requirements. To achieve this, the security model was designed with a number of design goals in mind:

The most important requirement is *interoperability* between products from different vendors, different security policies, and different security technologies. In addition, the security model should provide as much *transparency* as possible and *abstract* security-unaware applications from the complexities of the underlying security system, while at the same time allowing security-aware applications to enforce more specific and fine-grained policies. Administrators should be presented with a consolidated view of the system to make security management easier. Moreover, *flexibility* is needed to allow the use of many different security mechanisms and policy types. To achieve *portability*, the security model needs to be segregated as much as possible from underlying mechanisms, and the application needs to be isolated as much as possible from the security system. Also, consistent *integration* with preexisting security infrastructure is an important feature. The security architecture should not impose any *scalability* restrictions and should facilitate security administration for large systems by clustering principals and objects into groups and domains. A security system is only trustworthy if there is *assurance* that all security policies are automatically enforced correctly on all actions (i.e., malicious objects should not be able to bypass the security system). *Simplicity* of the security model helps reduce the number of potential bugs and, thus, improves the security system's reliability.

Realistically, the security architecture cannot live up to all of the design goals introduced, partly because some goals conflict with others (e.g., interoperability and flexibility, or flexibility and assurance), and partly because some design goals (e.g., assurance, performance, and scalability) depend on the quality of the particular implementation and underlying security mechanisms.

The CORBA security architecture consists of a number of functional components that provide applications with security. Figure 3.21 illustrates

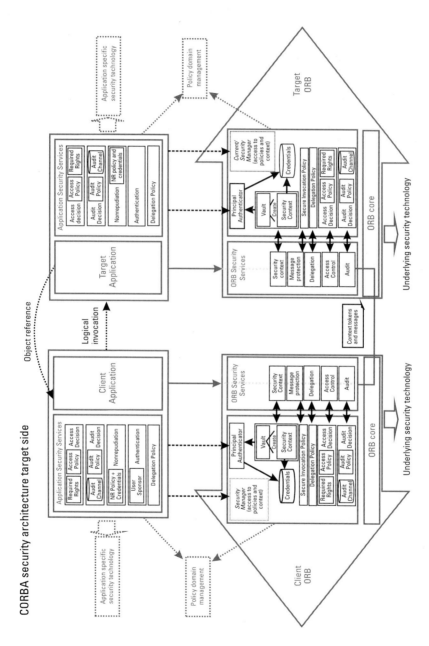

Figure 3.21 Overview of the CORBA security architecture.

the location of the main conceptual components and the interactions between them.

The CORBA security model is based on active subjects who can invoke operations on passive target objects, but only by going through a security enforcement component. Subjects and objects are called principals if they have a unique identity and a number of credentials. The security policies are based on these credentials. To allow security policy administration of large-scale systems, the security model clusters objects into domains and principals into groups (or roles, clearances, etc.). Security policies can be enforced either by the ORB or by the application. The ORB layer security policy is transparently enforced by the ORB for both security-aware and security-unaware objects, while application layer security policies are enforced only by security-aware application objects, which may have their own security requirements.

In distributed object systems like CORBA, an invoked object may in turn call other objects to perform parts of the task. This process is called *privilege delegation*—the act whereby one principal authorizes another to use its identity or privileges, perhaps with restrictions. Delegation policies on the ORB layer and on the application layer specify how credentials can be delegated (on the initiator side), how they should be passed on (on the intermediate side), and which credentials should be accepted (on the target side).

The *principal authentication* component allows principals to authenticate themselves to the system and create their personal credentials. Human users normally provide their claimed identity and authentication information to a *user sponsor* (inside or outside the object system), which then calls the principal authenticator on their behalf. The principal authentication policy is implicitly enforced by the underlying authentication mechanism: Based on the provided security information, the authentication mechanism will decide which privileges to put into the credentials object.

When the principal wants to securely invoke a remote object, the CORBA security system needs to associate its credentials with the communications, and, in particular, transfer the credentials securely to the remote peer. This is done as part of the *security association establishment* process. In the model, the security context is set up by the Vault object. The Current and SecurityManager objects are just another way of accessing information associated with the active security context. To support the protocol exchanges necessary for security context establishment, the IIOP requires a number of enhancements that are part of the SECIOP.

Now the actual invocation request can be sent across the network. *Message protection* policies on the ORB and application layer, as well as in the object reference, specify the minimum quality of protection that needs to be applied to each message (i.e., origin authentication, integrity, confidentiality) and the maximum supported quality of protection. The security system can combine these individual policies to arrive at the effective quality of protection that has to be applied to the message. Message protection (i.e., encryption) is enforced by the `SecurityContext` object.

Access control is about controlling which target methods each principal is allowed to invoke. The access policy object returns the granted rights for the calling principal, whereas the required rights object returns the required rights to call the target. By comparing both, the access decision object can find out if the invocation should be granted or not and enforce this decision. Again, access control can be done transparently on the ORB layer and more fine-grained on the application layer.

Security audit is the process of recording details about security-relevant events in an audit log (also called audit channel). The audit decision object triggers the generation of a log entry based on the audit policy, which specifies the types of events that have to be recorded. As before, audit policies can be defined on the ORB layer and on the application layer.

Nonrepudiation is about generating irrefutable evidence (proof of origin and receipt) that prevents participants in an action from convincingly denying their responsibility. This proof consists of cryptographically signed evidence tokens, which are generated by the `NRCredentials` object. This object can only be used by applications (i.e., not transparently), so that generated evidence can be undeniably attributed to a particular object or principal.

CORBA was originally developed for Intranet-centric environments with central administration. The growing use of CORBA on the Internet changed the situation: Now there was mutual suspicion between participants who operated on potentially hostile underlying platforms, and central security administration was not possible anymore in these highly dispersed systems. In addition, several firewalls were in the communications path between the client and target, which broke CORBA's end-to-end security model. There are several attempts to support firewalls within the CORBA security model (tunneling, IIOP layer proxy, TCP layer proxy). To meet the growing demand for Internet security, the lightweight security protocol SSL/TLS was also incorporated below the model as an alternative transport mechanism (SSLIOP), which can be used instead of SECIOP.

It became clear early on in the specification life cycle that most CORBA security vendors were not able or willing to implement products that cover the whole range of specified security features. Consequently, it was decided by the OMG that not all products needed to support all the facilities defined in the CORBA security specification. Instead, two levels of conformance were specified: For the first level of conformance (level 1 security), it is mandatory to provide a specified set of security features during object invocation for security-unaware applications and to allow security-aware applications to enforce their own access control and auditing. The second conformance level (level 2 security) includes a wider range of security facilities and allows applications to control the security provided at object invocation. It also includes security policy administration. Further conformance options specify nonrepudiation, security service replaceability, and secure interoperability.

3.8   Further Reading: Books on CORBA Security

At the time of this writing, there is not a lot of material available on the CORBA security architecture. The only book entirely dedicated to CORBA security is by Bob Blakley [22], which provides a lightly written introduction to the key concepts and components with entertaining examples. On the other extreme, the most detailed and technical work on the CORBA security architecture can, of course, be found in the CORBA security services specification [1]. However, the current version is 420 pages long, not very intuitively structured, and not easy to read. In addition, Hartman's book [23] covers CORBA and EJB in the more applied context of enterprise security. Dieter Gollmann's book on computer security [24] also contains a very brief and abstract introduction to CORBA security. Other introductory articles on CORBA security include those mentioned in the reference list [25–28].

References

[1]   OMG, *CORBA Security Service Version 1.8* (Draft Adopted Revision), 2000.

[2]   Lang, U., and R. Schreiner, "Flexibility and Interoperability in CORBA Security," *Electronic Notes in Theoretical Computer Science,* Vol. 26, Elsevier, Amsterdam, The Netherlands, February 2000.

[3]   Lang, U., R. Schreiner, and D. Gollmann, "Cryptography and Middleware Security," p. 408 in Qing, S. et al., "Information and Communications Security, Third International Conference, Xi'an, China, Nov. 2001, Proceedings," *ICICS' 2001, Lecture Notes in Computer Science No. 2229,* Springer Verlag, Berlin, 2001.

[4]   Adams, C., *The Simple Public-Key GSS-API Mechanism,* RFC 2024, 1996.

[5]   Kohl, J., and C. Neuman, *The Kerberos Network Authentication Service Version 5*, RFC 1510, 1993.

[6]   Bellovin, S., and M. Merrit, "Limitations of the Kerberos Authentication System," *In Proceedings of the USENIX Winter '91 Conference*, Dallas, 1991, pp. 253–267.

[7]   Parker, T., and D. Pinkas, *SESAME V4—Overview*, 1995, www.esat.kuleuven.ac.be/cusic/sesame/doc-txt/overview.txt.

[8]   Kaijser, P., and D. Pinkas, "SESAME: The Solution to Security for Open Distributed Systems," *Computer Communications*, Vol. 17, Issue 7, July 10, 1994, pp. 501–518.

[9]   Dierks, T., and C. Allen, *The TLS Protocol Version 1.0*, RFC 2246, January 1999.

[10] OMG, *Security Domain Membership Management Service* (final submission), July 2001.

[11] OMG, *Revised Public Key Infrastructure Submission*, February 1999.

[12] U.S. Department of Defense, "DoD Trusted Computer System Evaluation Criteria" (The *Orange Book*), DOD 5200.28-STD, 1985.

[13] Commission of the European Communities, *Information Technology Security Evaluation Criteria (ITSEC)*, Version 1.2, 1991.

[14] International Organization for Standardization, *ISO IS 15408 Common Criteria Version 2.1*, CCIB, ISO/IEC, 2000.

[15] Bell, D., and L. LaPadula, "Secure Computer Systems: Mathematical Foundations and Model," *MITRE Report MTR 2547*, Volume 2, 1973.

[16] Clark, D. R., and D. R. Wilson, "A Comparison of Commercial and Military Computer Security Policies," *Proceedings of the 1987 IEEE Symposium on Security and Privacy*, Oakland, CA, 1987, pp. 184–194.

[17] OMG, *Common Secure Interoperability Version 2 Specification*, 2001.

[18] Chapman, D., and E. Zwicky, *Building Internet Firewalls*, Sebastopol, CA: O'Reilly, 1995.

[19] Linn, J., *Generic Security Service Application Program Interface Version 2, Update 1* (IETF RFC 2743), 2000.

[20] Farrell, S., and R. Housley, *An Internet Attribute Certificate Profile for Authorization*, (IETF ID PKIXAC), 2000.

[21] Zakinthinos, Aris, *On the Composition of Security Properties*, University of Toronto and Centre for Communications Systems Research, University of Cambridge, 1996.

[22] Blakley, B., *CORBA Security—An Introduction to Safe Computing with Objects*, Reading, MA: Addison-Wesley, 2000.

[23]   Hartman, B. et al., *Enterprise Security with EJB and CORBA*, New York: Wiley/OMG, 2001.

[24]   Gollmann, D., *Computer Security*, Chichester, UK: Wiley & Sons, 1999.

[25]   Lang, U., *CORBA Security—Security Aspects*, M.Sc. dissertation, University of London, 1997, http://www.ulrichlang.com/research.html.

[26]   Lang, U., and R. Schreiner, *Schutz und Trutz, Sicherheit in CORBA-Basierten Systemen, iX*, October 1998.

[27]   Fairthorne, B., and B. Blakley, *Introduction to CORBA Security*, DOCSec 1997, Baltimore, MD, April 1997, http://cgi.omg.org/homepages/docsec/1997/index.html.

[28]   OMG Security SIG, OMG SecSIG Web site, http://www.omg.org/security, 2001.

4

Getting Started with MICOSec

4.1 Introduction

In Chapters 1 through 3, you learned the mainly theoretical concepts of CORBA and its security architecture. In Chapters 4 through 6, you will see how CORBA security is used in practice. This chapter is mainly concerned with installing and configuring the ORB and the security services, while Chapters 5 and 6 discuss using the application-facing CORBA security interfaces. Finally, Chapter 7 illustrates the use of CORBA security for security-unaware applications.

All examples in this book use MICOSec, our CORBA security services implementation, which was originally developed for research purposes and is "freely" available. It was designed to be used with the MICO ORB, which is also available as "free" software. MICOSec is implemented as conformant to the specification as possible, so most of the examples given here should also work with other compliant implementations.

4.2 Free Software

MICOSec is available as "free" software, a term coined in 1983 when Richard Stallman announced his plans to develop a freely available version of UNIX called GNU [1]. GNU is a nested acronym that stands for "GNU's Not Unix." The GNU project later defined "free software" as a matter of liberty (like free speech), not price (such as free beer) [2]. In practice, however,

129

this distinction is fuzzy. After all, free software in the sense of free speech is automatically available to anyone, and so it does not make sense to charge for it. As a consequence, free software is in practice more like both free speech and free beer.

According to the GNU philosophy, users are allowed to run, copy, distribute, study, change, and improve the software, as long as they adhere to the licensing conditions laid out in the GNU General Public License (GPL) or the GNU Library General Public License (LGPL).

In general, these licenses ensure that everybody who modifies free software makes the resulting code available under the same licensing conditions. This even applies if free software is used in conjunction with any software that is licensed differently (e.g., commercial products). Any software that contains components that are under the GPL must in turn be freely available. In a way, GPL software infects all other software it is used with, and therefore this property is often referred to as viral.

The main difference between the GPL and LGPL is that the latter allows the use of free software within proprietary programs if they are distinct software libraries that are just linked into the application. In this case, the application source code can remain unpublished, as long as the library source code is freely available. Products that adhere to the GPL or the LGPL are also approved by the Open Source Initiative (OSI) as *OSI Certified Open Source Software* [3].

4.3   MICO

The MICO project [4] was founded in December 1996 by Arno Puder to provide a freely available and fully compliant implementation of the CORBA 2.3 standard [5]. The original name MICO extended to Mini CORBA, but when MICO became bigger, the name did not seem appropriate anymore, and, consequently, the meaning was changed to MICO Is CORBA (inspired by the acronym GNU). In June 1999, MICO was branded CORBA compliant by the OpenGroup [6], effective proof that OpenSource [3] can produce quality software.

4.3.1   Overview

MICO is a C++ based CORBA 2.3 compliant ORB implementation. It does not require any proprietary libraries and tools to be compiled or executed. Resulting from its academic origin, the MICO implementation has a clear and modular design even for implementing internals to ensure easy extensibility.

Recent versions of MICO (v 2.3.x) include the following features (see [7] for details):

- Compiler for IDL to C++ mapping;

- Dynamic invocation interface (DII) and dynamic skeleton interface (DSI);

- IR and graphical interface repository browser that allows invoking of arbitrary methods on arbitrary interfaces;

- IIOP as native protocol (ORB prepared for multiprotocol support);

- Object adapters: POA and BOA, including all activation modes, support for object migration, and the implementation repository. The BOA can load object implementations into clients at run-time using loadable modules;

- Support for secure communication and authentication using SSL;

- Interceptor interfaces;

- Objects by value (OBV);

- Support for nested method invocations;

- The type "Any" offers an interface for inserting and extracting constructed types that were not known at compile time (dynamic "Any" is also supported);

- Support for using MICO from within X11 applications (Xt, Qt, and Gtk) and Tcl/Tk;

- CORBA services, such as interoperable naming service, trading service, event service, relationship service, property service, time service.

4.3.2 Installation

To get you started, this section briefly describes how MICO is compiled and configured. Please refer to the MICO user's guides [7, 8] for more detailed information if you encounter any problems during installation. The MICO sources can be obtained from the MICO Web site.[1]

---

1. http://www.mico.org

To compile MICO on UNIX platforms, you will need recent versions of GNU make (v3.7 or newer) and GNU gcc (v2.95 or newer) and its matching libg++ library. As an alternative compiler, egcs (v1.x) can be used. To compile the graphical user interface, flex (v.2.5.2 or newer) and bison (v1.22 or newer) are also required. And if you would like to compile the graphical interface repository browser, then you will also need Sun's JDK (v1.1.5) and the parser generator JavaCUP (v0.10g).

Note that you will not need to install MICO separate from MICOSec, as the latter comes as a bundle with all the MICO sources. Section 4.4.2 explains in detail how MICOSec is installed. If you would like to install the plain nonsecurity MICO, just unzip the archive for UNIX into a subdirectory called mico using the following command:

```
gzip -dc mico-<version>.tar.gz | tar xf -
```

Next, the MICO Makefile needs to be configured for your system. This can be done automatically by running configure with the proper command line options. The option -help gives you an overview of all supported command line options (alternatively, a list of all options can also be found in the MICO user's guide [7]).

```
cd mico
./configure <options>
```

To compile and install the programs and libraries, use gmake:

```
gmake
gmake install
```

MICO has been tested on Solaris, FreeBSD, AIX, Linux, Digital Unix, HP-UX, Ultrix, and Microsoft Windows NT or 95, but it should also run on a number of other platforms (see [7] for details).

4.4 MICOSec

MICOSec [9, 10] was originally developed as part of a research project in the telecommunications sector to analyze the viability of middleware security concepts. All sources and code examples are freely available from the MICO-Sec Web site, http://www.micosec.org.

4.4.1 Overview

MICOSec is a level 2 conformant implementation of the *CORBA Security Services v1.7 draft* [11]. This draft standard was used because the specification's last official version v1.2 [12] was considered outdated at the time the implementation was planned.

MICO was chosen as the underlying ORB because it is freely available, C++ based, very modular and clearly structured, and thus easily extensible.

One of the original requirements was to analyze CORBA security in open systems such as the Internet. As a result, SSL, which is widely used on the Internet today, was selected as a basic security mechanism for MICOSec. This allowed the use of MICO's built-in proprietary SSLIOP implementation, which is largely based on OpenSSL. Further, it was decided not to support Kerberos, a security mechanism hardly used outside the United States.

The current version of the Open Source version of MICOSec supports:

- All features of MICO, including the POA;
- Security for security-aware and security-unaware applications;
- IIOP for unprotected communications;
- SSLIOP, based on SSLv3 with all cryptographic algorithms supported by OpenSSL;
- Extensible attributes for X.509 certificates;
- Policies for secure invocation;
- Principal authentication;
- Message protection;
- Extended level 1 interfaces;
- Security domain mapping with domain membership management;
- Domain-based access control;
- Domain-based auditing with storage of audit records in various channels (file, UNIX syslog, PostgreSQL database).

The functionality for principal authentication, security context establishment, and message protection are all based on SSL. Although MICOSec was implemented as conformant as possible, it was decided to add some extensions in order to leverage all features of SSL. In particular, extra security attributes were added to make the content of X.509 certificates available to clients and targets. This way, it is possible to query security attributes such as the organization or certification authority of a certificate, which facilitates the specification of security policies in large systems with many users. With the standard security attributes, only the identity of the principal can be obtained. Note that most features of MICOSec are limited by the functionality provided by SSL (e.g., SSL only supports authentication of the target or mutual authentication, while authentication of the client is not possible).

Both the access control and audit policies use the domain names provided by the object domain mapping to express targets (see Section 6.6). Both the access control and audit service were implemented using request-level interceptors, which are initialized when the application is launched. This way, the services are automatically called by the ORB whenever a message arrives, which allows access control and auditing for security-unaware applications. In addition, it is possible for security-aware applications to define their own application-level access control and audit policy. The audit records can be recorded into a flat file, UNIX `syslog`, or a relational database (e.g., the Open Source database `PostgreSQL`).

4.4.2 Installation

MICOSec is released as a complete distribution package that includes MICO. The security additions are directly included in the MICO sources, so that the set-up procedure is almost unchanged from the standard MICO installation. Therefore, the installation of MICOSec is relatively trivial once you manage to successfully install MICO. The main additional installation hurdle is to set up various prerequisites necessary for running MICOSec.

MICOSec has been tested on FreeBSD, Linux, and Solaris (Sparc, with `gcc`). If you have problems building it, check if your platform is supported by the standard MICO (Version 2.3.6). Some platforms and compilers have problems with MICO's C++-code in general. MICOSec compiles best with the GNU C++ compiler `gcc` (v 2.95.2).

4.4.2.1 Prerequisites

Before MICOSec can be installed, all underlying mechanisms have to be set up: The `OpenSSL` library, which is used by MICOSec as its basic underlying security mechanism for authentication and message protection, and the `PostgreSQL` database, which can be used to store audit records.

Note that MICOSec also requires `flex` (v.2.5.2 or newer), which is preinstalled on most UNIX systems. If you do not have it, you can download it from the MICO Web site.

OpenSSL

First, the Open Source SSL library OpenSSL (v0.9.6b or later[2]) has to be installed, if it is not already preinstalled on the used operating system. Older versions of OpenSSL are not supported, in particular OpenSSL's predecessor SSLeay, which cannot be used with MICOSec.

To install OpenSSL on UNIX platforms, you will need Perl 5, an ANSI C compatible compiler, and a supported Unix operating system. Decompress the archive into a temporary directory with the following command:

```
tar xvfz openssl-<version>.tar.gz
```

Then run the following commands inside the openssl-<version> directory:

```
./config
make
make test
make install
```

This will build and install OpenSSL in the default location /usr/local/ssl. If you want to install it anywhere else, run config like this:

```
./config —prefix=/usr/local —
openssldir=/usr/local/openssl
```

If any of these steps fails, see the installation notes that come with the OpenSSL archive. The installation notes also specify several options to ./config (or ./Configure) to customize the build, such as noshared, which prevents the creation of shared libraries.

Note that sometimes problems are caused by incorrect or mixed installations of OpenSSL and SSLeay. In this case, the MICOSec build fails with obscure errors. Make sure you only use the latest version of OpenSSL.

It takes several rather complex commands to generate X.509 certificate files (.pem files). We will discuss how to generate such X.509 certificates in Section 5.4.2. For the moment, you can simply use the pregenerated

---

2. Available from http://www.openssl.org

certificates that come with the examples shipped as part of the MICOSec distribution.

PostgreSQL

If the audit data should be stored in a SQL database, MICOSec also requires the Open Source object-relational database management system (DBMS) PostgreSQL to be installed. PostgreSQL (in the past also called Postgres95) is derived from the Postgres package (developed at the University of California at Berkeley). With more than a decade of development behind it, PostgreSQL is the most advanced Open Source database available today, offering multiversion concurrency control and supporting almost all SQL constructs.

Before you start, create a PostgreSQL superuser account, from which the server will run. The PostgreSQL superuser owns the PostgreSQL binaries and database files. As the database superuser, all protection mechanisms may be bypassed and any data accessed arbitrarily. In addition, the PostgreSQL superuser is allowed to execute some support programs that are generally not available to all users. Note that the PostgreSQL superuser is not the same as the UNIX root superuser. For security reasons, the PostgreSQL superuser should be a separate, unprivileged account [i.e., with a nonzero user identifier (UID)]. If PostgreSQL is preinstalled on your UNIX system, then this account will be called postgres. The following installation instructions assume this account name for the PostgreSQL superuser account, and that you are logged on as the PostgreSQL superuser.

The code and installation information for UNIX platforms can be found at the PostgreSQL Web site.[3] Download the archive from there and uncompress it with the following command:

```
tar xvfz postgresql-<version>.tar.gz
```

Next, configure the makefiles for your system. It is this step at which you can specify your actual installation path for the build process and make choices about what gets installed. Inside the postgresql-<version> directory, change into the src subdirectory and type:

```
./configure <options>
```

3. Available from http://www.postgresql.org

For a complete list of options, use:

```
./configure -help
```

Note that building PostgreSQL requires GNU make. It will not work with other make programs. To compile the program, simply type:

```
gmake
```

Next, install the PostgreSQL executable files and libraries. You should complete this step as the user that you want the installed executables to be owned by (this does not have to be the same as the database superuser).

```
gmake install
```

Now that the database binaries have been installed, a new database installation (i.e., the working data files) needs to be created. To do this, you must log in to your PostgreSQL superuser account postgres (for security reasons, it will not work as root).

```
mkdir /usr/local/pgsql/data
chown postgres /usr/local/pgsql/data
su - postgres
/usr/local/pgsql/bin/initdb -D /usr/local/pgsql/data
```

The -D option ("directory") specifies the location where the data will be stored. Make sure that the superuser account can write to the directory before starting initdb.

You can now start the database server by running the following command:

```
/usr/local/pgsql/bin/postmaster -D
/usr/local/pgsql/data
```

If you prefer to put the server in the background instead, then use the following command:

```
nohup /usr/local/pgsql/bin/postmaster -D
/usr/local/pgsql/data \
</dev/null >>server.log 2>>1 &
```

The `postmaster` is the process that acts as a clearinghouse for requests to the `PostgreSQL` system (i.e., keeps track of any system errors and communication between the back-end processes). The MICOSec audit function will connect to the `postmaster`.

If you generated shared libraries, tell your system how to find the new shared libraries. How to do this varies between different platforms. One method is to set the environment variable `LD_LIBRARY_PATH`:

- For `sh`, `ksh`, `bash`, `zsh`:

```
LD_LIBRARY_PATH=/usr/local/pgsql/lib
export LD_LIBRARY_PATH
```

- For `csh` or `tcsh`:

```
setenv LD_LIBRARY_PATH /usr/local/pgsql/lib
```

- Another option is to use `export` as follows:

```
export PGLIB=/usr/local/pgsql/lib
export PGDATA=/usr/local/pgsql/data
```

Finally, to make sure the MICOSec run-time can use the `PostgreSQL` database, you need to set up some environment variables. First, you probably want to include `/usr/local/pgsql/bin` (or equivalent) into your PATH. To do this, add the following to your shell start-up file, such as `~/.bash_profile` (or `/etc/profile`, if you want it to affect every user):

```
PATH=$PATH:/usr/local/pgsql/bin
export PATH
```

Once the server is running, a new database can be created. Assume you want to create a database named `auditdb`, then you can do this with the following command:

```
createdb auditdb
```

Database names must have an alphabetic first character and are limited to 31 characters in length. Note that not every user has authorization to become a database administrator. If `PostgreSQL` refuses to create databases for you, then the administrator needs to grant you permission to create databases.

You can also test the newly built server. To do so, run the regression tests (`/src/test/regress`), a test suite to verify that `PostgreSQL` runs properly on your machine.

`PostgreSQL` is a complex piece of software, and its installation and configuration is not simple. If you run into any problems, the best option is to consult the *PostgreSQL Administrator's Guide* [13] for more information.

4.4.2.2 MICOSec Installation

Once these components have been installed, proceed with the standard MICO installation from the sources as already described in the MICO installation guidelines. The first step is to configure MICOSec. This is done the same way as for MICO, but using the following MICOSec-specific options:

- First, make sure that MICO's SSL support is enabled by using the following configuration parameter, where `<path>` should point to the directory where `OpenSSL` has been installed:

  `—with-ssl=<path>`

- By default, MICOSec only supports the CORBA Security level 1 interfaces. To enable the interfaces for CORBA Security level 2, the following configuration parameter has to be included:

  `—enable-csl2`

- By default, MICOSec only supports the storage of audit records in a flat file and UNIX `syslog`. To configure MICOSec to use `PostgreSQL` as the SQL database, add the following parameter, where `<path>` points to the directory where `PostgreSQL` has been installed:

  `—with-pgsql=<path>`

The following example illustrates how MICOSec is configured for level 2 security with `OpenSSL` and `PostgreSQL` installed at the specified locations:

```
./configure —with-ssl=/usr/local/ssl —enable-csl2
— with-pgsql=/usr/local/pgsql
```

After the correct configuration has been set up, MICOSec is built with MICO as described in Section 4.3.2. To compile and install the programs and libraries for UNIX, use `gmake`:

```
gmake
gmake install
```

4.5 Summary

MICOSec is a freely available level 2 conformant implementation of the *CORBA Security Services v1.7* draft [11]. The main property of free software is that the source code is publicly available, and that users are allowed to run, copy, distribute, study, change, and improve the software, as long as they make their code also freely available.

The current version of MICOSec supports all features of MICO, including the POA, security for security-aware and security-unaware applications, IIOP for unprotected communications, SSLIOP based on `OpenSSL` SSLv3, extensible attributes for X.509 certificates, policies for secure invocation and auditing, principal authentication, message protection, extended level 1 interfaces, and auditing with storage of audit records in various channels (file, UNIX `syslog`, SQL database).

MICOSec comes as a combined package with the MICO ORB, which is a freely available and fully compliant C++ implementation of the CORBA 2.3 standard that runs on a number of different platforms. The installation of MICOSec is almost identical to the MICO installation as both products reside within the same source tree. Before MICOSec can be compiled, `OpenSSL` and the `PostgreSQL` database system (both are also available as Open Source) have to be installed.

4.6 Further Reading on MICO and MICOSec

The *MICOSec User's Guide* [10] includes essentially the same information on how to install MICOSec as this chapter. For more information on how to install and configure `OpenSSL`, look at the respective installation notes that come with the source code. For more information on the `PostgreSQL` database, consult the *PostgreSQL Administrator's Guide* [13].

More information on installation and configuration of the MICO ORB can be found in *MICO—An Open Source CORBA 2.3 Implementation*

[8]. It explains how to install and use MICO, documents all features, and includes a tutorial.

References

[1]   Stallman, R., *GNU Project Announcement*, http://www.gnu.org/gnu/initial-announcement.html, 1983.

[2]   Free Software Foundation, *GNU Project—What Is Free Software?*, http://www.gnu.org/philosophy/free-sw.html, 2000.

[3]   OpenSource Initiative, *The Open Source Definition (Version 1.7)*, http://www.opensource.org/osd.html, 2000.

[4]   MICO Project, *MICO is CORBA*, http://www.mico.org, 2001.

[5]   OMG, *The Common Object Request Broker: Architecture and Specification*, 2000.

[6]   OpenGroup, *CORBA Open Brand*, Press Release, June 7, 1999, http://www.opengroup.org/press/7jun99_a.htm.

[7]   MICO Project, *MICO—An Open Source CORBA 2.3 Implementation, Version 2.3.5*, http://www.mico.org, 2000.

[8]   Römer, K., A. Puder, and F. Pilhofer, *MICO is CORBA, An Open Source CORBA 2.3 Implementation*, San Francisco, CA: Morgan Kaufman Publishers, 1999.

[9]   Schreiner, R., and U. Lang, *MICOSec User's Guide*, ObjectSecurity Ltd., 2000, http://www.micosec.org.

[10] Schreiner, R., and U. Lang, *MICOSec Reference Manual*, ObjectSecurity Ltd., 2000, http://www.micosec.org.

[11] OMG, *CORBA Security Services Draft Version 1.7*, 1999.

[12] OMG, *CORBA Security Services Specification*, 1998.

[13] PostgreSQL Development Team (editor: Thomas Lockhart), *PostgreSQL Administrator's Guide*, PostgreSQL, Inc., 2000, http://www.sk.postgresql.org/devel-corner/docs/admin/admin.html.

5

Security Level 1

5.1 Introduction

This hands-on chapter will illustrate how you can use CORBA security conformance level 1 interfaces in your applications. To do this, we will extend the Bank example introduced in Chapter 1 to use the interfaces available at security level 1.

To support level 1, CORBA security implementations have to implement a subset of the full CORBA security functionality, as well as the application-facing level 1 interface that allows applications to access security attributes from the current security context. Most of the functionality is provided automatically on the ORB layer (i.e., for security-unaware applications). Remember that security-unaware means that security policies can be enforced for an application without requiring any modifications to the application code. The application is simply linked together with a security-enhanced ORB and then becomes secure, provided an adequate policy is in place.

Level 2 includes a wider range of security facilities and allows applications to control the security provided at object invocation. It also includes security policy administration. Further conformance options specify nonrepudiation, security service replaceability, and secure interoperability. The use of level 2 interfaces will be discussed in detail in Chapter 6, and Chapter 7 will cover using CORBA security for security-unaware applications.

At first glance, supporting level 1 interfaces does not make sense for security services implementations that, like MICOSec, also support all level 2 interfaces. After all, anyone who just wants to use the level 1 functionality could use the relevant subset of the level 2 interfaces. But this is not the case, as the level 1 interface is not a strict subset of the level 2 interfaces. The level 1 interface allows applications to access security attributes directly, whereas, the level 2 security introduces the more complex concept of credentials, which contain the security attributes. Consequently, the level 1 interface is easier to handle.

MICOSec supports both the level 1 and level 2 interfaces, but since the level 1 implementation is internally based on the level 2 functionality, it is not possible to enable the level 1 interface without enabling the level 2 interfaces at the same time. MICOSec also implements the functionality required for level 1 or level 2 security-unaware applications, which will be discussed in Chapter 7. This chapter only covers the use of MICOSec's level 1 interfaces.

5.2   Level 1 Functionality

According to the CORBA security services specification [1], level 1 conformant implementations should automatically provide the following functionality to security-unaware applications:

- Principal authentication inside *or* outside the object system;
- Secure invocation between client and target object (including unilateral authentication, integrity and/or confidentiality) on the ORB layer or outside the object system;
- Simple delegation of client security attributes to targets, depending on the supported CSI level;
- ORB-enforced access control checks (with support for domains and roles), but no support for administration;
- Auditing of security-relevant system events (but not by object invocation).

The specification neither mandates any internal interfaces nor gives any implementation guidelines; therefore, security services implementers are free to provide the specified functionality whichever way they prefer. For MICOSec, it was most elegant to reuse the richer implementation that had already been developed in conformance with the security-unaware level 2 functionality (see Chapter 7), instead of producing a separate implementation for level 1.

This chapter is mainly concerned with the application-facing level 1 interfaces, which security-aware applications can use to retrieve security attributes directly. These attributes can then be used to enforce application-specific security policies (e.g., access control based on the application state). The level 1 interface allows access to security options and attribute details.

5.3 Level 1 Interface

Despite its limited functionality, the level 1 interface is very useful for many real-world applications. It provides a very simple and convenient way to obtain the peer attributes directly from the Current object (i.e., the security context), without having to bother with the handling of more complex level 2 Credentials objects. In practice, this is exactly the functionality that is often needed to allow applications to enforce their own specific security requirements.

To access security attributes, the CORBA security services specification only defines a single level 1 interface:

```
module SecurityLevel1 {
   interface Current : CORBA::Current {
      Security::AttributeList get_attributes (
            in Security::AttributeTypeList attributes
         );
      };
   };
```

IDL 1: CORBA security level 1 interface.

The operation Current::get_attributes allows an application (i.e., target-side implementation) to obtain the security attributes of the client on whose behalf it is operating. The most common example of such attributes would be the identity of the calling principal. The application can then use these attributes to control access to its functions or data and to log security-relevant events.

However, the simple level 1 interface cannot support the reverse flow of security attributes (i.e., allow client-side applications to obtain security attributes of the target-side). This is because in a servant there is a secure association with exactly one client at a time. On the client side, however, there may be more than one association with different servers, and within the specified interface, there is no way to select for which association the attributes should be obtained.

However, in SSL-based applications, authentication of the client by the target is not enough. The client also wants to know which target it is communicating with. To solve this problem, MICOSec level 1 introduces an additional operation `Current::get_target_attributes`. This interface allows client applications to select a secure association to a particular target by its IOR, which is passed to MICOSec as an additional parameter. `Current::get_target_attributes` returns the attributes of the target corresponding to the association selected by the IOR. The following IDL shows the extended MICOSec level 1 interface that includes the operation `Current::get_target_attributes`:

```
module SecurityLevel1 {
    interface Current : CORBA::Current {
        Security::AttributeList get_attributes (
            in Security::AttributeTypeList attributes
            );
        Security::AttributeList get_target_attributes (
            in Security::AttributeTypeList attributes,
            in Object obj
            );
    };
};
```

IDL 2: Extended MICOSec security level 1 interface.

Note that this operation is a nonstandard addition to the *CORBA security services specification v1.7*, which should only be used if portability across different ORB and security services products is not an issue.

Despite its elegance and simplicity, this extended level 1 interface does not fully conform to the conventions set out in the specification. *CORBA Security Services Version 1.5* and above distinguish between `SecurityManager` and `Current`. The `Current` object is only used to obtain thread-specific security information in servants, whereas the `SecurityManager` object, which is associated with the process as a whole, should be used in the server and client. So, in order to conform to this, the operation should more appropriately be moved to `SecurityManager::get_target_ attributes`, but for simplicity reasons it was decided to keep it in `Current`.

5.4　A Security-Aware Bank Application Example

Remember the bank application example introduced in Chapter 1, in which a bank server should maintain accounts for its clients. Bank account objects offer the following three operations: `deposit` a certain amount of money, `withdraw` a certain amount of money, and an operation `balance` that returns the current account balance. The state of an account object consists of the current balance. The `Bank` interface provides an operation to create new bank accounts. The following IDL file `account.idl` captures that functionality:

```
interface Account {
    void deposit( in unsigned long amount );
    void withdraw( in unsigned long amount );
    long balance();
};

interface Bank {
    Account create ();
};
```

IDL 3: `account.idl`.

In this chapter, we will use the standard level 1 interface to allow the target-side bank application to check the caller's security attributes. In real-world applications, these attributes could be used to authenticate the legitimate owner of a bank account. In addition, we use the extended MICOSec interface on the client-side to retrieve the target's security attributes. This way, clients could ensure that they are talking to the real bank and not to some malicious application that exports the same interface.

For the sake of simplicity, our example does not contain any policy evaluation code. Instead, it simply outputs the remote peer's security attributes on the console, so you can visually check the security attributes of the remote peer. This illustrates well how level 1 functionality can be used in practice, but at the same time it keeps the example code short and simple.

5.4.1　Building and Running the Example

Build the programs in the `/demo/security/l1` subdirectory using the provided `Makefile`. The building process works exactly the same way as for the standard MICO demo programs. First, the IDL file is translated with the arguments for using the POA, then the client and target code is compiled and linked. From a linking perspective, the MICOSec code is part of the MICO library. Therefore, it is not necessary to link an additional security library in order to security-enhance the demo program.

Running programs differs slightly from normal MICO use, because security-enhanced applications need to access the security information from the OpenSSL certificate and key files. But since there are no application-facing level 1 interfaces that allow the specification of the certificate and key file names from within the application source code, an alternate approach is necessary—the certificate and key files have to be provided at the command line when the application is launched.

The bank server is started with the following small shell script called rss:

```
#!/bin/sh
ADDR=ssl:inet:'uname -n':12456
./server -ORBIIOPAddr $ADDR -ORBSSLcert
ServerCert.pem -ORBSSLkey ServerKey.pem
-ORBSSLverify 0
```

Shell script: Server shell script.

The command line arguments are the same as the MICO SSL options and will be described in more detail in Section 5.4.2:

- ORBIIOPAddr defines the socket of the server. Note that the "ssl:" in the ADDR variable tells the ORB to bind to a SSL-socket waiting for SSLIOP requests instead of to a simple TCP-socket serving IIOP.
- ORBSSLcert defines the X.509 certificate to use.
- ORBSSLkey defines the OpenSSL private key pair file.
- ORBSSLverify defines the depth of the certificate verification path.

When running the shell script, you should get the following console output:

```
% ./rss
Start Bank server
narrow to SecurityLevel1Server::Current
Running.
```

The first line just informs you that the ORB is being initialized. After that, a reference to `SecurityLevel1Server::Current` is created, and the `Bank` object is initiated. The target-side bank implementation is now ready to process requests.

The client is started in the script `rcs`:

```
#!/bin/sh
ADDR=ssl:inet:'uname -n':12456
./client  -ORBBindAddr $ADDR -ORBSSLcert
ClientCert.pem -ORBSSLkey ClientKey.pem
-ORBSSLverify 0
```

Shell script: Client shell script.

The command line arguments are the same, except for the additional argument `ORBBindAddr`, which specifies the socket to which the client should connect.

You can now start the client in another command line window and see from the console output how MICOSec level 1 provides the client application with the security attribute of the target-side X.509 certificate. The underlying SSL protocol makes sure that all this information is authentic:

```
% ./rcs
SSL verify error: self signed certificate
Received 1 attributes
1 2 /C=UK/ST=Server State/L=Cambridge/O=ObjectSecurity
Ltd./OU=RD/CN=Server Test
deposit - 700
withdraw - 450
Balance is 250.
%
```

Shell script: Client-side screen output.

Note that the example certificates that come with the `Bank` code cause the `OpenSSL` implementation to generate an error message, which states that the certificate is self-signed. This is correct because no X.509 certificate authority file was given in the example. SSL will be discussed in more detail in Section 5.4.2.

The application then indicates that it retrieved a security attribute and outputs the server identity before it invokes any operations on the server.

On the server side, we get the following output when the shell script is started:

```
% ./rss
Start Bank server
narrow to SecurityLevel1Server::Current
Running.
SSL verify error: self signed certificate
Received  1 attributes
1 2 /C=UK/ST=Client State/L=Cambridge/O=ObjectSecurity
Ltd./OU=RD/CN=Client Test
```

Shell script: Target-side screen output.

As expected, the target application (in fact, the `withdraw` servant implementation, as you will see in Section 5.4.3) outputs the calling remote peer identity.

In real-world applications, the obtained security information could be used both on the client and target side for simple application-level access control and auditing. For example, the target application could compare the caller identity to a list of clients that are allowed to invoke a particular operation.

Note that in this example, both client and server use a MICO-specific bind mechanism, which is not conformant to the CORBA 2.3 standard. Of course, it would also be possible to instead use the standard mechanism that involves IORs. For the sake of simplicity, it is assumed in our example that both client and server are executed on the same host. If you would like to run the client on machines other than the server host, simply replace the `ADDR` value (which now points to the local host described by its hostname) with the appropriate hostname of the server.

5.4.2   SSL and X.509 Certificates

MICOSec reuses MICO's built-in SSL support for its authentication and message protection. MICO, in turn, uses the underlying `OpenSSL` library, which provides the actual SSL functionality.

The SSL protocol was originally developed by Netscape, mainly to protect World Wide Web traffic. The IETF developed a draft on Transport Layer Security (TLS), which is almost identical with SSL Version 3, and so the protocol is now known as SSL/TLS [2]. Logically, SSL resides above the TCP layer and below the middleware layer. From a CORBA perspective, SSL is just another transport layer below IIOP.

SSL keeps its own session state, which includes information related to cryptographic algorithms, such as a session identifier, the specification of the cipher suite, shared secret keys, and certificates. The actual protocol is divided into two components, the handshake layer and the record layer. The handshake layer is concerned with the negotiation of the used cipher suite, with establishing the necessary keying material, and with authentication, while the record layer is responsible for the encryption. For authentication, SSL uses X.509 identity certificates, which have to be countersigned by a certificate authority (CA) to bind the identity to a cryptographic public key pair.

In essence, the SSL handshake works like this: The client-side SSL library initiates the protocol by sending a message with a random number, a list of suggested ciphers ordered according to the client's preference, and maybe a suggested compression algorithm. The target-side SSL library then selects one cipher suite from the suggested list (including the algorithm for key exchange, encryption, and hashing) and sends it back to the client, together with its certificate and another random number. In this message, the target can also request a certificate from the client. Once the used cipher has been defined, the client generates a random number (PreMasterSecret), which is transformed into keying information (MasterSecret) using the random numbers generated by the client and target. The PreMasterSecret is then securely sent to the target, using the key management algorithm specified in the selected cipher suite and the target's public key. With this information, the target can also generate the MasterSecret, so that both parties possess a shared secret key. After that, all traffic can be encrypted by the record layer protocol.

To configure the SSL protocol, MICOSec uses the following MICO SSL command line arguments:

- `ORBSSLcert <certificate file>`

  This command line option specifies the file that holds the X.509 certificate for the launched client or target. `OpenSSL` files use the extension `.pem` for key and certificate files. This argument defaults to `default.pem`.

- `ORBSSLkey <key file>`

  This option specifies the `.pem` file that holds the key pair for the launched client or target. It defaults to the same file as the certificate file.

- `ORBSSLcipher <colon separated list of preferred ciphers>`

  This parameter can be used to specify the ciphers that the launched client or target is willing to support. If it is not specified, then an implementation-specific default policy is used instead, which depends on the cryptographic functions supported by the specific implementation, as well as on cryptography export regulations and patents in some countries.

  Commonly used cipher suites include: NULL-MD5, RC4-MD5, EXP-RC4-MD5, IDEA-CBC-MD5, RC2-CBC-MD5, EXP-RC2-CBC-MD5, DES-CBC-MD5, DES-CBC-SHA, DES-CBC3-MD5, DES-CBC3-SHA, and DES-CFB-M1.

- `ORBSSLverify <verify depth>`

  If this parameter is specified, then the peer must supply a valid certificate; otherwise the connection setup will fail. `<verify depth>` specifies how many hops of the chain of certification authorities should be checked. By default, the validity of the peer certificate is not checked.

- `ORBSSLCAfile <CAfilename>`

  This argument specifies the `.pem` file that holds the certificates of CAs.

- `ORBSSLCApath <CA pathname>`

  This parameter can point to the directory that contains `.pem` files holding certificates of CAs. It defaults to `/usr/local/ssl/certs`.

SSL 3.0 generally provides excellent security against eavesdropping and other passive attacks, but in CORBA it is sometimes possible to obtain useful information even from encrypted requests, for example, if different objects are bound to different ports. The most dangerous passive attack is counting the byte length of an encrypted request: A CORBA operation with fixed-length data types normally has the same length during transmission, even when encrypted, so a passive attacker might be able to derive the operation invoked just by counting the transmitted bytes. Traffic padding would be necessary to counter these so-called traffic analysis attacks, but the CORBA security services explicitly do not support that.

Academic research also identified a number of active attacks that could be carried out against some SSL implementations, most notably the *change cipher spec-dropping attack* (also called *cipher suite rollback attack*), in which an attacker forces a connection into a weaker cipher than necessary by editing the cipher suite list in the handshake protocol. Another attack along these lines involves spoofing of the key exchange algorithm (*also called key exchange algorithm rollback attack*), to force the use of a weaker key exchange algorithm [3]. Both attacks can easily be prevented if the chosen cipher suite is checked by the application, or if both client and target are configured to accept only a single matching cipher suite.

You might have noticed two SSL-related error messages when you executed the example:

```
SSL verify error: self signed certificate
SSL verify error: Certificate has expired
```

This is not an error, it simply states that the certificates that come with the example are not trustworthy because they are self-signed and expired. Also, no X.509 certificate authority file was specified in the example.

For secure real-world CORBA applications, you should generate trustworthy keys, certificates, and certificate authorities. To generate a new private key file (by default saved as `privkey.pem`), you can use the following `openssl` parameters:

```
openssl genrsa  (RSA private key), or
openssl gendsa  (DSA private key)
```

Next, you will need to get a certificate from a CA, which binds your identity to the generated private key. To do that, you will need to generate a *certificate signing request* and send it to your certification authority. It will then have to sign it and return the corresponding certificate. To generate a request, use the following command:

```
openssl req -new -key privkey.pem -out req.pem
```

Now, `req.pem` can be sent to the certificate authority, where the certificate can be generated with the command:

```
openssl ca -in req.pem -out newcert.pem
```

More information on key generation and CAs can be found in the OpenSSL manual pages [4]. Note that the configuration and maintenance of a full X.509 PKI is a complex and laborious task, which is not directly related to CORBA security and, thus, will not be covered in detail in this book. Please consult specialist literature if you would like to know more about PKIs and X.509 [5].

5.4.3 The Target

In this section, we will describe how the level 1 security interface is used from the target-side example bank application. The nonsecured version of the account example was introduced in Chapter 1, and we will now compare it with this security-enhanced version to see the main differences. If you are not familiar with the (nonsecure) Bank example, you can also consult the MICO documentation [6] for more details on how MICO and the demo programs work. The nonsecured account example is part of the standard MICO distribution (in the subdirectory demo). The source code and IDL files for the security-enhanced level 1 version can be found in the subdirectory demo/security/l1.

Like any other CORBA application, the target-side implementation comes in two logical parts. The *server* part is used to launch the application and the ORB, whereas the implementation of the actual functionality behind the target object's interface resides in the *servant*.

We will first look at the server source code. First, level 1 security requires an instance of the SecurityCurrent object (named security-current and seccur in this example) for further use:

```
/*
 * A Bank factory that creates Account objects
 */
```

```
#include "account.h"

CORBA::ORB_var orb;
CORBA::Object_var securitycurrent;
SecurityLevel1::Current_var seccur;
```

The first part of the main server function remains unchanged from the non-secured version. It initializes the ORB, the POA, and the POA manager.

```
int main (int argc, char *argv[])
{

    /*
     * Initialize the ORB
     */

    cout < "Start Bank server\n";
    orb = CORBA::ORB_init (argc, argv, "mico-local-orb");

    /*
     * Obtain a reference to the RootPOA and its Manager
     */

    PortableServer::POA_var poa;
    CORBA::Object_var poaobj =
        orb-> resolve_initial_references ("RootPOA");
    poa = PortableServer::POA::_narrow (poaobj);
    PortableServer::POAManager_var mgr =
        poa-> the_POAManager();
```

The only security-related addition to the server involves getting a pointer to the `Current` object. Remember that `Current` contains the security information associated with a particular session. The main purpose of the level 1 interface is to allow applications to retrieve this information from the application layer. This `Current` object is obtained by resolving an initial reference called `SecurityCurrent` and then narrowing it to the `SecurityLevel1::Current` type.

Note that, in this example, the resulting `SecurityLevel1:: Current` object instance `seccur` is declared global. This is done to avoid

having to resolve and narrow the initial reference each time a servant that uses
seccur is called. Using local variables would be cleaner for real-world applications, but in this example the goal is to keep the code as simple as possible. The
reference to SecurityCurrent is obtained and stored in seccur as follows:

```
/*
 * Get SecurityCurrent
 */

securitycurrent = orb->
    resolve_initial_references ("SecurityCurrent");
cout << "narrow to SecurityLevel1Server::Current\n";

seccur =
    SecurityLevel1::Current::_narrow(securitycurrent);

assert (!CORBA::is_nil (seccur));
```

The rest of the server source code remains unchanged from the unsecured
version. It first creates a new Bank object instance and tells the POA to activate it. After that, it activates the POA manager and the ORB. The target-
side bank application is now ready to receive requests from the client side.

```
/*
 * Create a Bank
 */

Bank_impl * micocash = new Bank_impl;

/*
 * Activate the Bank
 */

PortableServer::ObjectId_var oid =
    poa->activate_object (micocash);

/*
 * Activate the POA and start serving requests
 */
```

```
printf ("Running.\n");
mgr->activate ();
orb->run ();

/*
 * Shutdown (never reached)
 */

poa->destroy (TRUE, TRUE);
delete micocash;

return 0;
}
```

The servant implements the functionality of the Account and Bank interfaces. Account provides operations to deposit an amount of money, withdraw money, and query the balance of the account. The first code fragment
initializes the Account object with a zero balance:

```
/*
 * Implementation of the Account
 */

class Account_impl : virtual public POA_Account
{
public:
  Account_impl ();

  void deposit (CORBA::ULong);
  void withdraw (CORBA::ULong);
  CORBA::Long balance ();

private:
  CORBA::Long bal;
};

Account_impl::Account_impl ()
{
  bal = 0;
}
```

In this example, only the `withdraw` servant implementation has been modified to use the operation `get_attributes` from the level 1 interfaces to obtain a list of security attributes from `SecurityCurrent`.

The `get_attributes` operation needs to be provided with an *attribute list*, a standard data structure defined in the specification. As an input parameter, this list specifies which attributes should be obtained. After the call has completed, the list contains the obtained security attributes as an output parameter. This list variable is generated and populated in several steps. First, an attribute variable `at` has to be created. Next, the types of the security attributes that should be obtained have to specified. The CORBA security services specification defines a number of such standard security *attribute families* and *attribute types* in the file `security.idl`. On top of these, MICOSec defines a number of additional attributes that provide access to SSL-specific information, such as the X.509 properties of the remote principal, as well as environmental information, such as the hostname of the remote peer.

Security attributes consist of an *attribute value* (e.g., the principals identity) and a *defining authority*. To achieve maximum flexibility, the attribute value content is of the type `Opaque` (i.e., a sequence of plain octets). However, this makes the portability of the attribute value interpretation practically impossible, because the meaning of the attribute content normally depends on the particular security mechanisms used. MICOSec returns as value a text string with the requested information, so it can easily be printed and processed with string functions. The `defining_authority` parameter is rarely used in practice. It describes who defined the associated attribute type (not who defined the information in the value). For all OMG-defined standard types, this parameter is empty.

In our example application, we only want to obtain a single attribute—the client access identity—which is identified by family <0:1> and attribute `AccessID` (a full list of available attributes is described in Section 5.4.4). So an attribute list of length one is created (named `at1`), into which the generated attribute `at` is stored.

To keep the code as short as possible, the example simply prints the content of the client's access identity on the standard output whenever the client withdraws money from his account. In practice, you would, of course, rather compare the retrieved attributes with the target's security policy and only grant the withdraw operation if they match. The example could easily be extended to grant or deny access based on the caller's `AccessID` and an access control list. For simplicity, the `deposit` and `balance` operations remain unchanged.

```
void Account_impl::deposit (CORBA::ULong amount)
{
  bal += amount;
}

void Account_impl::withdraw (CORBA::ULong amount)
{
  Security::ExtensibleFamily fam;
  fam.family_definer = 0;
  fam.family = 1;
  Security::AttributeType at;
  at.attribute_family = fam;
  at.attribute_type = Security::AccessId;
  Security::AttributeTypeList atl;
  atl.length(1);
  atl[0]=at;

  Security::AttributeList_var al =
    seccur->get_attributes( atl );

    cout
      << "Received  "
      << (*al).length()
      << " attributes\n";

  for ( int ctr = 0; ctr (*al).length(); ctr++) {

  cout
      <<(*al)[ctr].attribute_type.attribute_family
         .family
      << " "
      << (*al)[ctr].attribute_type.attribute_type << " "
      << &(*al)[ctr].value[0] << " "
      << &(*al)[ctr].defining_authority[0]
      << endl;
  }
  bal -= amount;
}
```

```
CORBA::Long Account_impl::balance ()
{
  return bal;
}
```

The Bank interface implementation remains unchanged from the nonsecured version. Its create operation creates and activates new bank accounts, and returns a reference to the new bank account to the caller.

```
/*
 * Implementation of the Bank
 */

class Bank_impl : virtual public POA_Bank
{
public:
  Account_ptr create ();
};

Account_ptr Bank_impl::create ()
{

  /*
   * Create a new account (which is never deleted)
   */

    Account_impl * ai = new Account_impl;
  /*
   * Obtain a reference using _this.
   * This implicitly activates the
   * account servant
   * (the RootPOA, which is the object's _default_POA,
   * has the IMPLICIT_ACTIVATION policy)
   */
  Account_ptr aref = ai->_this ();
  assert (!CORBA::is_nil (aref));
  /*
   * Return the reference
   */
  return aref;
}
```

5.4.4   Security Attributes

In the previous section, you learned how the requested security attributes are specified and passed to MICOSec. This section provides an overview of the kinds of security attribute types available to the application.

5.4.4.1 Standard Attributes

The CORBA security services specification defines two standard attribute families and their types, which are summarized in Table 5.1.

The standard CORBASec attribute families define separate identities for different security aspects of the peer, such as an identity for access control (`AccessId`), auditing (`AuditId`), accounting (`AccountingId`), and non-repudiation (`NonRepudiationId`). SSL, the security mechanism used by MICOSec, supports only one identity, so all these different CORBASec identities are set to this single identity described by the X.509 certificate.

The `Public` attribute caters to unauthenticated or anonymous peers. If a peer did not authenticate itself to CORBASec, it has only this single default attribute.

Some security mechanisms, such as SESAME, support roles and groups, and so CORBASec defines the attribute types `PrimaryGroupId` to

Table 5.1
Standard Attribute Types

Family 0—Identity Attributes	
1	AuditId
2	AccountingId
3	NonRepudiationId

Family 1—Privilege Attributes	
1	Public
2	AccessId
3	PrimaryGroupId
4	GroupId
5	Role
6	AttributeSet
7	Clearance
8	Capability

obtain this information. According to the specification, SSL-based implementations do not use this attribute type and the security enforcement is only based on the peer's identity. In practice this is not sufficient, and so MICOSec maps the peer's X.509 organization's unit to the `Primary GroupId` attribute type.

`GroupId`, `Role`, `AttributeSet`, `Clearance`, and `Capability` cannot be supported with SSL as the underlying security mechanism, and are therefore not used by MICOSec.

5.4.4.2   MICOSec Attribute Families 10 and 11

On top of that, MICOSec uses the extended attribute family identifiers 10 and 11 for its security attributes. Note that the CORBA security services specification mandates that custom attribute families are to be located above family 7. The MICOSec specific attribute type family 10 allows access to the content of X.509 certificates, whereas family 11 provides other information from the current security context.

In family 10, MICOSec specifies extended attributes for X.509 certificates, because the standard CORBASec attribute families do not fit well to SSL. On the one hand, they define attribute types that are not supported by SSL and, on the other hand, they do not specify enough attributes to retrieve all the information from X.509 certificates. For example, there is no attribute type to obtain the certification authority that has issued a certificate. The MICOSec specific attribute family 10 provides a better way of processing information directly from X.509 certificates (see Table 5.2).

The `X509Subject` attribute identifies the entity associated with the certificate. The entity is called *subject* and is described by a nonempty distinguished name.

`X509Issuer` identifies the CA that has signed and issued the certificate. This field contains the issuer's distinguished name.

As already described in Section 5.4.2, a cipher suite is chosen during the SSL security context establishment between client and server. The chosen cipher suite depends on the cryptographic algorithms supported by client and server and on the session establishment policy set by the user. The designation of the cipher suite used for the security context can be obtained with the attribute type `X509Cipher`.

The attributes 4 through 9 are just shortcuts to the different parts of the distinguished name of the subject. Attributes 10 through 15 specify the analogous parts for the issuer. Instead of obtaining the distinguished name and then parsing the returned string for the desired information, they can be

Table 5.2
MICOSec Extended Attribute Family 10

Family 10—X.509 Certificate Attributes		
1	X509Subject	Subject's identity
2	X509Issuer	Issuer's identity
3	X509Cipher	Cipher suite
4	X509Subject_CN	Subject's designation
5	X509Subject_C	Subject's country
6	X509Subject_L	Subject's city
7	X509Subject_ST	Subject's state
8	X509Subject_O	Subject's organization
9	X509Subject_OU	Subject's organization unit
10	X509Issuer_CN	Issuer's designation
11	X509Issuer_C	Issuer's country
12	X509Issuer_L	Issuer's city
13	X509Issuer_ST	Issuer's state
14	X509Issuer_O	Issuer's organization
15	X509Issuer_OU	Issuer's organization unit

retrieved directly using these attribute types. The MICOSec specific family 11 provides additional access to useful low-level information (see Table 5.3). AuthMethod specifies the authentication method used. In MICOSec, it always returns the string "ssl". In addition, the remote peer's complete network address, the hostname, and the port number can be obtained with the attribute PeerAddress.

Table 5.3
MICOSec Extended Attribute Family 11

Family 11—MICOSec Low Level Attributes	
1	AuthMethod
2	PeerAddress

5.4.5   The Client

In our bank account example, the client would also like to know which target it is talking to. For many real-world applications this is a security requirement, because it can prevent confidential data (e.g., transaction numbers in a banking environment) from being accidentally sent to the wrong server.

To obtain the target's security attributes through level 1 interfaces, the client example application has been modified similarly to the target's `withdraw` operation. But first, the client has to instantiate its `Security Level1::Current` object:

```
#include "account.h"

CORBA::ORB_var orb;
CORBA::Object_var securitycurrent;
SecurityLevel1::Current_var seccur;

int main (int argc, char *argv[])
{
  CORBA::ORB_var orb =
      CORBA::ORB_init (argc, argv, "mico-local-orb");
```

Just like on the target side, the client first has to obtain a reference to the `SecurityLevel1::Current` object by resolving and narrowing the initial reference to `SecurityCurrent`:

```
/*
 *  Get SecurityCurrent
 */

securitycurrent = orb->resolve_initial_references
    ("SecurityCurrent");

/*
 *  Narrow to SecurityLevel1Server::Current
 */

seccur =
    SecurityLevel1::Current::_narrow(securitycurrent);
assert (!CORBA::is_nil (seccur));
```

Next, the client has to bind to the `Bank` server. In the nonsecure bank example, this binding just sets up a network connection between the client and server. In this security-enhanced version, it also sets up an SSL connection between the peers and exchanges the security information associated with the security session, such as keys and certificates.

```
/*
 * Connect to the Bank
 */

CORBA::Object_var obj = orb->bind ("IDL:Bank:1.0");

if (CORBA::is_nil (obj)) {
  printf ("oops: bind to Bank failed\n");
  exit (1);
}

Bank_var bank = Bank::_narrow (obj);
assert (!CORBA::is_nil (bank));
```

Before operations on the `Bank` and `Account` interfaces can be called, the requested security attributes have to be specified within an attribute list variable. This is done in exactly the same way as on the target side. First, the family and the security attribute `AccessId` have to be created and specified. After that, a security attribute list is created and populated with the generated security attribute.

```
/*
 * Get and print attributes of server
 */

Security::ExtensibleFamily fam;
fam.family_definer = 0;
fam.family = 1;
Security::AttributeType at;
at.attribute_family = fam;
at.attribute_type = Security::AccessId;
Security::AttributeTypeList atl;
atl.length(1);
atl[0]=at;
```

Now the target security attributes can be obtained by calling `get_target_` `attributes` on the `SecurityLevel1::Current` object. This noncon-formant addition to the level 1 interface is necessary to provide clients with a means to select the security context from which the security attributes should be retrieved. The context can be selected by passing the corresponding server IOR as an additional argument:

```
Security::AttributeList_var al =
    seccur-get_target_attributes( atl, bank );
```

The retrieved security attribute list content is then printed on the standard output in exactly the same way as on the target side example:

```
cout << "Received "
        << (*al).length()
        << " attributes\n";

for ( int ctr = 0; ctr (*al).length(); ctr++) {

    cout
      <<(*al)[ctr].attribute_type.
        attribute_family.family
    << " "
    << (*al)[ctr].attribute_type.attribute_type
    << " "
    << &(*al)[ctr].value[0]
    << " "
    << &(*al)[ctr].defining_authority[0]
    << endl;
}
```

After the client has connected to the target-side bank application and obtained its security attributes through the level 1 security interface, it creates a new `Account` object and uses the operations to deposit and withdraw cash and query the account balance.

```
/*
 * Open an account
 */
```

```
Account_var account = bank->create ();
if (CORBA::is_nil (account)) {
  printf ("oops: account is nil\n");
  exit (1);
}

/*
 * Deposit and withdraw some money
 */

cout << "deposit - 700\n";
account->deposit (700);
cout << "withdraw - 450\n";
account->withdraw (450);
printf ("Balance is %ld.\n", account-balance ());
return 0;
}
```

5.5 Implementation Overview and Conformance

Level 1 security is not spectacular; for a lot of its basic functionality, it just reuses the functionality provided by SSL. In particular, level 1 does not give applications direct control over the ORB layer security functionality—the application-facing interfaces only allow applications to retrieve security attributes from the current security context.

Principal authentication is done by the underlying SSL library. Remember that, for level 1, the X.509 identity certificate has to be supplied from the command line when the client and target applications are launched. This certificate is then simply passed on to the `PrincipalAuthenticator` object, and a `Credentials` object with the corresponding security attributes is attached to the `Current` security context. In our level 1 implementation, the `Credentials` object simply contains the whole certificate as the identity. The peer object can then verify the authenticity of the claimed identity from the certificate. The SSL handshake automatically makes sure that the certificates are transferred securely across the SSL connection.

MICOSec's security context establishment is automatically launched when the ORB tries to bind to a remote object. Remember from the general discussion in Chapter 3 that the security service is connected to the ORB through so-called interceptor interfaces. For most security mechanisms, MICOSec would therefore have to listen to the request-level interceptor provided by ORB and establish a security context whenever the ORB opens a new connection.

But SSL is not a normal security mechanism; for the ORB, SSL is just another transport mechanism, very much like TCP. So whenever the ORB wants to connect to a target object for which there is not already a security context in place, it calls the underlying SSL library to establish an SSL connection between the SSL sockets on both sides.

In our case, however, MICO already comes with its own SSL support, so instead of calling its own SSL implementation, MICOSec communicates with `OpenSSL` through the interfaces provided by MICO. Also, the SSL configuration option initializes MICO in such a way that it automatically uses SSL instead of plain TCP to connect to the target. It is due to this particular feature that, in this implementation, the actual flow of the certificates between the ORB and MICOSec is the reverse of what you would expect. It is MICO rather than MICOSec that provides the hook to SSL. Initially, the certificate is provided to MICO's SSL implementation when the application is launched. MICOSec then retrieves them from MICO's SSL hooks when the SSL security context has been set up. It then puts them into its `SecurityLevel1::Current` object in order to make it available to the application layer. So MICOSec essentially piggybacks onto MICO's SSL support, which is an elegant design feature. From the application layer, the whole SSL handshake is fully transparent.

SSL supports a number of different options for peer authentication (in particular, unilateral or mutual authentication), as well as a number of different cryptographic algorithms and key lengths for peer authentication and message protection. This flexibility was a design requirement for its original use as a security protocol for World Wide Web traffic, because export regulations in the 1990s prohibited the export and use of some cryptographic algorithms in a number of countries. Consequently, there were many SSL implementations with differing cipher suites, and so the negotiation of the used ciphers was necessary for each connection. The preferred order of cipher suites and which cipher suite is selected depends on the particular underlying hardware and software platform on the client and target side. On the other hand, the administrator can specify which algorithms should be available for negotiation and in which order of preference.

Using the negotiated cipher suite, as well as the loaded keys and certificates, SSL authenticates the peer and encrypts all network traffic associated with the established SSL context to protect it against unauthorized disclosure and modification. This is done automatically on the transport layer. All MICOSec level 1 has to do is retrieve information about the used ciphers and the involved identities from the SSL security context and put it into the SecurityLevel1::Current object, so that applications can retrieve all this information by using get_attributes and get_target_attributes. To achieve this, the authors of MICOSec decided to specify additional noncompliant security attributes for this X.509 specific information (see Section 5.4.4 for a complete list of security attributes). Our level 1 example only retrieves the access identity, but it could easily be extended to also retrieve the used cipher suite.

Level 1 conformance also mandates ORB-enforced access control (with support for domains and roles) and auditing (of security-relevant system events). But there are no application-facing interfaces, so this functionality has to be provided for security-unaware applications. The MICOSec implementation is also security level 2–conformant, and since the level 1 security-unaware functionality is a strict subset of level 2, it automatically conforms to the level 1 access control and audit requirements. A level 2 conformant security–unaware bank account example will be presented in Chapter 7.

Finally, the CSI part of the CORBA security services mandates the support of simple delegation for some underlying security mechanisms, in particular Kerberos (for CSI 1) and SESAME (for CSI 2). Simple delegation means that the target believes that it communicates with the client, although, in fact, it talks to the intermediate. This type of delegation is also called impersonation, because the intermediate object can use the caller's credentials as if they were its own. However, SSL is classified as CSI 0, for which no delegation support is required. There are a number of reasons why SSL cannot effectively support delegation: The delegation credential is essentially the X.509 certificate that binds the caller identity to the public key and, ultimately, to the SSL connection. If the intermediate is supposed to establish an SSL connection with exactly the same SSL properties as the one between the client and the intermediate, then it needs all keys and certificates of the client. In practice, this is totally against the basic rule that keys and certificates should not be shared between different parties, as this compromises accountability. Therefore, level 1 conformant simple delegation should not be supported by SSL-based CORBA security implementations (unless CSIv2-SAS is also supported, see Section 6.7.2.2).

5.6 Summary

This chapter describes how the CORBA security level 1 interfaces are used in practice. This is done by extending the `Bank` example introduced in Chapter 1 to use the interfaces available at security level 1.

Level 1 interfaces are not a strict subset of level 2 interfaces, but rather they provide a simpler way of accessing level 1 functionality. They allow applications to access security attributes directly, whereas level 2 security introduces the more complex concept of credentials, which contain the security attributes. Consequently, level 1 interfaces are easier to handle. The standard `SecurityLevel1::Current` interface only contains a single operation `get_attributes` for the target side. MICOSec provides an additional (nonstandard) operation `get_target_attributes` that allows the client side to retrieve the security attributes of a secure association with a particular target (specified by its IOR).

In addition to SSL-based authentication and message protection, level 1 security also includes ORB layer access control and audit. However, level 1 does not mandate any application-facing interfaces for that, so this functionality will instead be described in Chapter 7. Also note that, in accordance with CSI level 0, delegation is not supported by SSL.

When level 1 applications are launched, they have to be provided with a number of command line arguments that specify SSL-specific information, such as key and certificate files, as well as the server socket. Certificates can be generated and signed with `OpenSSL`.

In addition to the standard CORBA Security attribute, MICOSec introduces a number of SSL-specific security attributes to access the X.509 certificate content and other low-level information, such as the remote socket. These attributes can be used to access security context information directly from the application layer.

MICOSec level 1 elegantly reuses MICO's SSL support, which simply treats SSL as an alternative underlying transport mechanism. MICOSec retrieves the security context information from MICO and includes it in the CORBA security context, so that it can be easily retrieved through the level 1 interface.

5.7 Further Reading

There is no literature on the actual use of CORBA security level 1 interfaces. ObjectSecurity's *MICOSec User's Guide* [7] is the only other documentation on MICO Security level 1, but it is kept very brief and does not contain any information that goes beyond what is described in this chapter. Some specification details on interfaces and conformance can be found in the CORBA security services specification [1]. However, the information related to level 1 is not very readable, is spread throughout the lengthy specification, and does not give any explicit use guidelines.

References

[1]   OMG, *CORBA Security Services Specification*, 1998.

[2]   Dierks, T., and C. Allen, RFC 2246: The TLS Protocol, Version 1.0, January 1999.

[3]   Wagner, D., and B. Schneier, *Analysis of the SSL 3.0 Protocol*, Second USENIX Workshop on Electronic Commerce Proceedings, USENIX Press, November 1996, pp. 29–40.

[4]   OpenSSL team, Miscellaneous Open SSL documents, http://www.openssl.org/docs.

[5]   SourceForge Initiative. OpenSource PKI Book, http://ospkibook.sourceforge.net/, 2001.

[6]   Römer, K., A. Puder, and F. Pilhofer, *MICO is CORBA, An Open Source CORBA 2.3 Implementation*, San Francisco, CA: Morgan Kaufman Publishers, 1999.

[7]   Schreiner, R., and U. Lang, *MICOSec User's Guide*, ObjectSecurity Ltd., 2000, http://www.micosec.org.

6

Security Level 2

6.1 Introduction

In this chapter, you will learn how to use the full set of MICOSec's security features from within your application. For demonstration purposes, we will again extend our `Bank` example application from Chapter 1 to access underlying CORBA security features, but this time we will use the CORBA security level 2 interfaces.

Level 2 security incorporates a wider range of security facilities than level 1 and allows applications to control the security provided at object invocation at a finer granularity. It also supports interfaces for security policy administration. To make all functionality accessible from the application layer, level 2 security provides a rich set of application-facing interfaces. Note that this chapter is only concerned with application-facing interfaces. The use of CORBA security for security-unaware applications will be covered in Chapter 7.

Level 2 security has many advantages, but these benefits come at a price. While level 2 functionality and interfaces are richer and more flexible than at level 1, they are, at the same time, more complex in their use. In particular, level 2 security associations are based on the more flexible concept of `Credentials` objects instead of `Current`. Credentials contain the security attributes of local and remote principals, as well as fine-grained security association policies. We will describe the level 2 credentials model in Section 6.3.

This chapter is structured into several subsections, which present the interfaces to the different security functions. This way, the use pattern for each functional component can be studied in isolation. In real-world applications, such an approach is not recommended because there are many interactions between functional components that need to be considered. For example, principal authentication is often a precondition for access control.

In the following section, we will briefly look at the CORBA security level 2 functionality before describing the level 2 security-enhanced client and server example programs.

6.2   Level 2 Functionality Overview

For level 2 conformance, the security services need to support extra functionality on top of level 1 functionality. As level 1 functionality has already been described in Section 5.2, we will only describe which additional features are required.

At the ORB layer (i.e., for both security-unaware and security-aware) applications) level 2 conformant products need to provide the following functionality:

- Principal authentication both inside *and* outside the object system;

- Additional secure invocation features, in particular, peer authentication and message protection at the ORB level;

- Further integrity options, such as replay/reorder protection (can be requested, but need not be supported by all implementations);

- Access control (`DomainAccessPolicy`) and selective auditing have to support a per-operation granularity.

In addition, application-facing interfaces can be used to control in more detail (i.e., from security-aware applications):

- *Secure invocation:* Applications must be able to choose the quality of protection of messages required, change the privileges in credentials, and choose which credentials are to be used for object invocations.

- *Delegation:* Applications should also be able to specify whether credentials are to be used only at the target (e.g., for access control), or whether they can also be delegated. The application can request (unspecific) "composite" delegation, and the target can obtain all credentials passed, provided all participating nodes support this.

For policy administration, all security policy types (except nonrepudiation) have to be supported, and the standardized policy management interfaces for each of the level 2 policies have to be implemented. ORBs and applications must be able to obtain the ORB layer security policies that apply to them. Applications must also be able to locate and use their application layer policies to make decisions about what security is needed or to get the information needed to enforce the policy.

Level 2 conformant applications have to implement all application interfaces (except nonrepudiation, which is optional), all security policy administration interfaces, and all administrator's interfaces. However, this does not automatically imply that all specified values of privilege attributes, delegation modes, and communications options have to be implemented. Instead, some of these interfaces may raise a CORBA::NO-IMPLEMENT exception. Raising this exception can, for example, be necessary in cases where the underlying security mechanism cannot support a particular attribute, mode, or option.

And as with level 1, CORBA security services implementations that conform to level 2 can optionally provide any other specified security features. Currently, only nonrepudiation is specified as an optional security feature. To conform to the nonrepudiation option, all interfaces for evidence generation and verification (NRCredentials) and the nonrepudiation policy (NRPolicy) have to be implemented. Additional conformance options specify security service replaceability and secure interoperability. But since these options describe the interface between the security service implementation and the ORB, and not the interface between the application and the security service, they are not described here.

6.3 Principal Authentication and Secure Association

CORBA security level 2 is based on the flexible concept of using Credentials objects to describe security properties of principals and fine-grained policies for secure associations.

In this section, we will introduce the first example application, which demonstrates the use of two main security functions; principal authentication establishes the credentials of the client and the target Bank application, while the setup of a *secure association* transfers these credentials to the remote side. All credentials information can be accessed from the application layer and used for security enforcement.

6.3.1   Principal Authentication

`Credentials` objects hold the security information related to local (and remote) principals, such as the principal's privilege and identity attributes. Application programmers can obtain these security attributes from the CORBA security implementation.

The specification defines several different types of `Credentials`[1]: `OwnCredentials`, `ReceivedCredentials`, and `TargetCredentials`. `OwnCredentials` are normally created during principal authentication, whereas `ReceivedCredentials` (at the target) and `TargetCredentials` (at the client) are generated when a secure association is set up. In addition, it is possible get a copy of existing `Credentials` objects or ask for a `Credentials` object via `Current/SecurityManager`.

6.3.1.1   Own Credentials

The `Credentials` object (which, for the sake of clarity, is called `OwnCredentials` throughout this book to prevent any confusion with other types of credentials) is normally generated during principal authentication and holds the local security attributes of the principal associated with an application. `OwnCredentials` denotes the first element of the `own_credentials` list in the `Credentials` object. Both client and target have their `OwnCredentials`. In MICOSec, `OwnCredentials` objects are created by the `PrincipalAuthenticator` object (which is described in Section 6.3.1.2) and can be used by the application to access information from its own X.509 certificate. Both client and target have their `OwnCredentials`, which are exchanged as they establish a secure association.

In addition to the specific privilege and identity attributes for the communicating parties, all `Credentials` objects also contain at least one unspecific attribute of type `Public` (see Figure 6.1). This generic attribute allows the specification of policies that apply to anyone to be specified in much the same way as policies based on other, more restricted, attributes.

In addition, `Credentials` objects contain a number of more technical (read-only) security attributes, such as:

- The *(invocation) credentials type*, which specifies if the credentials are own credentials, received credentials, or target credentials;

---

1. In fact, the terms refer to the values that can be taken by the `Credentials::credentials_type` attribute.

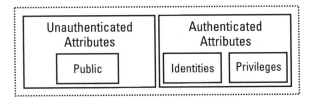

Figure 6.1 Credentials.

- The *authentication state*, which shows if the `Credentials` were successfully initialized during principal authentication or not, if the authentication has expired, or if a multistep authentication process is only partially completed;

- The *mechanism type*, which specifies both the mechanism supported for the secure association (e.g., Kerberos5) and the used cryptographic profiles (e.g., MD5_RSA);

- The *supported and required options* for both incoming and outgoing secure associations, such as integrity, confidentiality, replay detection, target or client authentication, as well as the supported delegation type.

Applications access `Credentials` through a number of operations, which provide the functionality to:

- Specify default security association options;

- Modify security attributes in the `Credentials`;

- Obtain information about the security attributes currently in the `Credentials`;

- Obtain information about a security feature for a given communications direction;

- Check if the `Credentials` are still valid, and refresh them if they have timed out;

- Create an exact copy of the `Credentials` object;

- Destroy a `Credentials` object.

6.3.1.2 Principal Authenticator

Before a CORBA client can securely invoke an object, it needs to establish its OwnCredentials by calling the PrincipalAuthenticator object (see Figure 6.2). The generated credentials are then associated with its principal by the CORBA security service, so that they can be automatically used to set up the secure association between communicating parties whenever that principal invokes an object.

PrincipalAuthenticator was originally only intended as the client-side interface behind a login client (or *user sponsor*). The login client would ask the user for his username and password and then invoke the authenticate operation on the PrincipalAuthenticator object to create its associated OwnCredentials. Alternatively, this login client could be implemented outside the CORBA system to allow for *single sign-on*. For example, the operating system user login could be tweaked to invoke PrincipalAuthenticator as part of the operating system login procedure, so that credentials could be generated automatically whenever a user logs in. With CORBA security services implementations based on traditional authentication mechanisms such as Kerberos or SESAME, the target application would get its OwnCredentials from some mechanism outside the CORBA system. However, in SSL-based implementations such as MICO-Sec, the target also needs to establish its OwnCredentials from a certificate by calling its PrincipalAuthenticator.

In addition to creating an OwnCredentials object, the Principal-Authenticator returns its authentication status, which can be successful, failed, or expired. In addition, the authentication process can be incomplete,

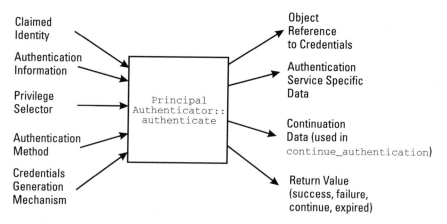

Figure 6.2 Principal authenticator interface.

as some authentication schemes, such as challenge-response mechanisms, may require more than one step to complete. To support these multistep authentication methods, `PrincipalAuthenticator` provides an operation `continue_authentication`, which has to be called until the authentication process has completed. SSL authentication is done within one step, so this functionality is not required in MICOSec.

6.3.2 Secure Association

The secure association is automatically set up by the CORBA security services when the client-side ORB tries to invoke an operation on a target for the first time. This process involves several parts. First, a handshake protocol is normally carried out to create an authenticated and protected communications channel based on some specified secure association options. On top of that, the client's `OwnCredentials` are securely transferred to the target-side ORB, and vice versa.

The exact inner workings of the secure association set-up depend on the nature of the underlying security mechanism. Normally, the ORB binding process establishes an unprotected network connection and then informs the security service implementation that a new binding exists. The security service then invokes its underlying security mechanism, which sets up the secure association on top of the existing network connection. While this is true for most security mechanisms, SSL-based implementations such as MICOSec require the integration of SSL as an alternative transport protocol into the ORB. This is because the SSL protocol already includes the set-up of a plain TCP/IP network connection and, thus, cannot be used on top of a preexisting network connection.

The transferred credentials are called `ReceivedCredentials` on the target-side and are available from the application layer through the `Current` interface. The client-side equivalent is called `TargetCredentials` and can be accessed through the `SecurityManager` interface (see Figure 6.3).

6.3.2.1 Received Credentials

This target-side object represents the secure association between the servant and its associated client. It contains the credentials of the authenticated client principal that made the invocation and, therefore, includes much of the same information as the client's `OwnCredentials` object, such as privilege attributes and identities.[2] `ReceivedCredentials` are used within servants

---

2. The `ReceivedCredentials` interface inherits from the `Credentials` interface.

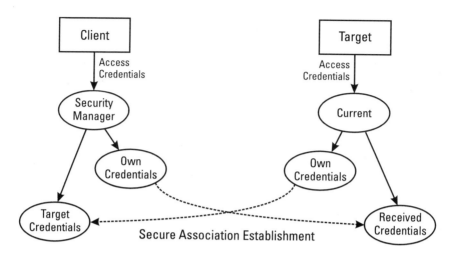

Figure 6.3 Received credentials and target credentials.

to obtain the attributes of the client that invoked the operation on the servant, the used association options, the delegation state of the remote principal, and the delegation mode of the `ReceivedCredentials`.

6.3.2.2 Target Credentials

This client-side object contains the target's security attributes and, thus, is the reverse of the `ReceivedCredentials` object. Target credentials can, for example, be used to check the target identity before any sensitive information is transferred to the target. Note that `TargetCredentials` may not be used for further invocations, while `ReceivedCredentials` can be delegated. If a client invokes an operation on different servers, then there are different target credentials for each server, so that on the client side each server is associated with its own set of target credentials.

6.3.3 Security-Aware Bank Example: Authentication and Secure Association

We will now extend the `Bank` example application from Chapter 1 to use the level 2 security interfaces, just as we did for the level 1 example in Section 5.4. It is based on the following IDL interface, which you are probably familiar with by now:

```
interface Account {
  void deposit( in unsigned long amount );
  void withdraw( in unsigned long amount );
  long balance();
};

interface Bank {
  Account create ();
};
```

IDL 1: `account.idl`

In the following sections, we will describe how the level 2 interfaces are used in MICOSec for principal authentication and secure association establishment. After that, subsequent sections will demonstrate using other security functionality components, such as access control and audit.

6.3.4 Building and Running the Example

The source code and IDL for this first level 2 security example can be found in the MICOSec directory `demo/security/tutorial`. It is built the same way as a normal POA-based MICO application, which has already been described. Additional libraries are not required in order to run this example.

The main difference to the level 1 security example described in Chapter 5 is that you do not need to provide any additional flags this time—the level 2 interfaces allow applications to take care of keys and certificates themselves. Executing the example is, therefore, easy; the client and server shell scripts only contain two lines each:

```
ADDR=ssl:inet:'uname -n':12456
./server -ORBIIOPAddr $ADDR
```

Server shell script

```
ADDR=ssl:inet:'uname -n':12456
./client -ORBIIOPAddr $ADDR
```

Client shell script

First, the server is started as usual by executing the shell script ./rss. It activates the target application, creates its OwnCredentials from an X.509 certificate, and prints the credentials information out on the console. Then it waits for invocations:

```
charon% ./rss
Start Bank server
Own credentials 1 attributes
family = 1 type = 2 /C=UK/ST=Server
State/L=Cambridge/O=ObjectSecurity
Ltd./OU=RD/CN=Server Test/Email=server@test
Running.
```

Next, the client is started by running ./rcs. Before it binds to the target, it also creates its OwnCredentials from the X.509 certificate and outputs them. It then outputs the target's identity attribute to demonstrate how security attributes can be obtained from the TargetCredentials:

```
charon% ./rcs
Own credentials 1 attributes
family = 1 type = 2 /C=UK/ST=Client
State/L=Cambridge/O=ObjectSecurity
Ltd./OU=RD/CN=Client Test/Email=client@test
Server credentials 2 attributes
family = 1 type = 2 /C=UK/ST=Server
State/L=Cambridge/O=ObjectSecurity Ltd./OU=
RD/CN=Server Test/Email=server@test
family = 11 type = 2
ssl:inet:charon.objectsecurity.com:12458
deposit - 700
withdraw - 450
Balance is 250.
charon%
```

When the withdraw operation is invoked, the server also outputs the identity attribute of the remote caller to demonstrate the use of the ReceivedCredentials:

```
Received credentials 1 attributes
1 2 /C=UK/ST=Client State/L=Cambridge/O=ObjectSecurity
Ltd./OU=RD/CN=Client Test/Email=client@test
```

The main difference between this example and the level 1 example in Chapter 5 is that this time no command line arguments are necessary. Both client and target get the file names of their X.509 certificates directly from the application source code.

Also note that this example uses an object reference (IOR) for the binding instead of the MICO-specific binding mechanism in CORBA. There are many ways to transfer object references from the target to client, including proprietary means, full-blown naming services, or other out-of-bound means of communication. For simplicity, this example transfers the object reference through a file. To do that, the target first stringifies the object reference and stores it in a file named `Bank.ref`. The client then reads that file and processes it back into an IOR object, which is then used to connect to the target. Note that this modification was only introduced to demonstrate the different binding methods throughout this book and has no relevance to the security interfaces as such.

6.3.5   The Target

The following minimal CORBA security level 2 target application demonstrates the use of objects for principal authentication and secure association establishment. The nonsecured version of the account example has already been introduced in Chapter 1, and we will now compare it with this security-enhanced version to see the main differences.

The target first obtains its process-specific initial object, the `SecurityManager`, and then creates its own `Credentials` objects using `PrincipalAuthenticator`. If the servant for the `withdraw` operation is called, an initial object `Current` is obtained from the ORB. This object is used to get the `ReceivedCredentials` object, which represents the secure association with the client that invoked the servant. The `get_attributes` operation is then used to obtain the attributes from the `Credentials` object.

Like any other CORBA application, the target-side implementation comes in two logical parts. The *server* part is used to launch the application and the ORB, whereas the implementation of the actual functionality behind the target object's interface resides in the *servant*. Both are analyzed in turn in the following subsections.

6.3.5.1   The Server

As in Chapter 5, we will look at the POA-based server source code first. The main nonsecurity-related difference to the level 1 example is that this example uses an IOR instead of the MICO-specific bind mechanism. First, level 2 security creates a pointer to a `SecurityManager` object:

```
#include <fstream.h>
#include "account.h"

CORBA::ORB_var orb;
CORBA::Object_var securitymanager;
SecurityLevel2::SecurityManager_var secman;
```

The first part of the main function remains unchanged from the unsecured version. It just initializes the ORB and the POA:

```
int main (int argc, char *argv[])
{

    cout   << "Start Bank server\n";
    orb = CORBA::ORB_init (argc, argv, "mico-local-orb");

    PortableServer::POA_var poa;
    CORBA::Object_var poaobj =
        orb-> resolve_initial_references ("RootPOA");
    poa = PortableServer::POA::_narrow (poaobj);
    PortableServer::POAManager_var mgr =
        poa-> the_POAManager ();
```

Next, the application needs to get the initial objects to reference the security service, just as it does for other services such as the naming service. These security-related objects are `SecurityManager` and `Current`.[3] `SecurityManager` is used to access capsule-specific security information associated

---

3. Until *CORBASec Version 1.5*, there was only `Current`. `SecurityManager` was introduced in Version 1.7.

with the ORB and its process, whereas `Current` contains security information related to thread-specific operation in servants.

In the main function of the server there are no separate threads; all security information is specific only to the ORB and the server process as a whole, so the correct initial reference is `SecurityManager`. The reference to this locally constrained object is obtained by calling the `resolve_initial_references` method on the ORB, followed by narrowing to the appropriate class. Operations on these objects are used to access the security information in the ORB, for example, to create credentials and get credentials or security policies.

```
securitymanager =
   orb-> resolve_initial_references
      ("SecurityManager");
assert (!CORBA::is_nil (securitymanager));

secman = SecurityLevel2::SecurityManager::
   _narrow(securitymanager);
assert (!CORBA::is_nil (secman));
```

Now the target's `OwnCredentials` object can be created from its associated X.509 certificate. This is done by calling the `authenticate` operation on the `PrincipalAuthenticator` object. But before that, we have to choose the correct authentication method:

```
Security::AuthenticationMethod our_method =
   (Security::AuthenticationMethod)
   SecurityLevel2::KeyCertCAPass;
```

The next step is to get an empty data structure for the authentication data from the `SecurityManager`. In the case of MICOSec, the X.509 certificate data needs to be provided to the underlying SSL implementation. Unfortunately, the integration of SSL into CORBA security services is underspecified, so the exact content of this data structure is implementation-specific and, thus, not portable. The used data structure therefore needs to be of the unspecific type `Any`:

```
Security::SSLKeyCertCAPass method_struct;

CORBA::Any* any_struct =
  secman -> get_method_data(our_method);
*any_struct >>= method_struct;
```

This structure must be filled with the correct information about the target's X.509 certificate: the file name of the server key, the file name of the server X.509 certificate, the file name of the certificate authority X.509 certificate, and the directory of the CA certificate (if it is not already supplied in CAfile). Also, it is possible to specify a password to unlock the key, but this is currently unsupported in MICOSec. The following code fragment shows how this is done:

```
method_struct.key = "ServerKey.pem";
method_struct.cert = "ServerCert.pem";
method_struct.CAfile = "list.pem";
method_struct.CAdir = "";
method_struct.pass = "";

CORBA::Any* out_any_struct;
out_any_struct = new CORBA::Any;
*out_any_struct <<= method_struct;
```

Remember that in our level 1 example application it was not possible to specify this information from within the application, so it had to be provided manually at the command line.

In the next step, we get a pointer to the PrincipalAuthenticator object from the SecurityManager:

```
SecurityLevel2::PrincipalAuthenticator_ptr pa =
  secman -> principal_authenticator();
```

Before the authenticate operation can be invoked, various arguments have to be specified:

- First, we can optionally select a user-specified cipher suite (e.g., IDEA-CBC-SHA) or use a default cipher.

- Next, the implementation-specific security name must be set to "ssl".

- The next line shows how the target can specify additional privileges, which it would like to be authenticated. This feature is currently not supported by MICOSec.

- The `creds` object stores the obtained `Credentials`. Often, the `OwnCredentials` attribute is used instead.

- The final two arguments are used for multiple-step authentication, such as challenge-response protocols. This is also not supported in the current version of MICOSec because, from the perspective of the CORBA security services, SSL completes in a single step.

The following code fragment shows how the arguments are defined:

```
const char* mechanism = "";

/* Optional:
 * const char* mechanism = "IDEA-CBC-SHA";
 */

const char* security_name = "ssl";

Security::AttributeList privileges;

SecurityLevel2::Credentials_ptr creds;

CORBA::Any* continuation_data;
CORBA::Any* auth_specific_data;
```

Now the `authenticate` operation can be called with all the specified arguments. It generates the target's own credentials and returns them in `Creds`. If something goes wrong, an exception is raised.

```
try
  {
      pa -> authenticate( our_method, mechanism,
      security_name, *out_any_struct,privileges,
      creds,continuation_data,auth_specific_data);
  }
  catch (...)
```

```
    {
        cout << "authentication failed" <<endl;
        delete out_any_struct;
        return 0;
    }
```

After successful authentication, the target's OwnCredentials are auto-matically entered as the first element of the SecurityManager's own credentials attribute list (own_credentials). This is how the Own-Credentials can be obtained from there:

```
SecurityLevel2::Credentials_ptr own_cred;
own_cred = (*(secman -> own_credentials())) [0];
```

We can then obtain the target's security attribute from the OwnCreden-tials object in a similar way as in the level 1 example. First, an attribute type list has to be instantiated and filled with the information about the requested attribute:

```
Security::ExtensibleFamily fam1;
fam1.family_definer = 0;
fam1.family = 1;
Security::AttributeType at1;
at1.attribute_family = fam1;
at1.attribute_type = Security::AccessId;
Security::AttributeTypeList atl1;
atl1.length(1);
atl1[0]=at1;
```

Then the Credentials object can be asked for the attributes.[4] In contrast to the level 1 server, get_attributes is now called on the Credentials object and not on SecurityManager. This allows CORBA security level 2 to obtain attributes from different Credentials to access different kinds of security information. To get the target's own attributes (which come from its X.509 certificate), the operation is invoked on the OwnCredentials

---

4. Some attribute types are not useful for the principal's own credentials. For example, the MICO-specific attribute "PeerAddress" does not make sense here, as there is no secure association with a remote peer at this point.

object. get_attributes returns a list of the desired attributes, which is printed out on the standard output just like in the level 1 example:

```
Security::AttributeList_var all =
    own_cred -> get_attributes(atl1);
cout << "Own credentials"
        << (*all).length() << " attributes\n";
for ( int ctr = 0; ctr (*all).length(); ctr++) {
    cout  << "family = "
            << (*all)[ctr].attribute_type.
            attribute_family.family
        << " "
        << "type = "
        << (*all)[ctr].attribute_type.attribute_type
        << " "
        << &(*all)[ctr].value[0]
        << " "
        << &(*all)[ctr].defining_authority[0]
        << endl;
    }
```

The application can get these credentials directly after it creates the initial object SecurityManager (i.e., before it activates the POA or receives any requests from clients). This is possible both in client and target applications and can, for example, be used to check the identity of the user principal.

The last part of the service remains unchanged. It creates a Bank object instance, registers it with the POA, writes the IOR to a file, and activates the ORB and the POA manager:

```
Bank_impl * micocash = new Bank_impl;

PortableServer::ObjectId_var oid =
    poa-> activate_object (micocash);

ofstream of ("Bank.ref");
CORBA::Object_var ref =
    poa-> id_to_reference (oid.in());
CORBA::String_var str =
    orb-> object_to_string (ref.in());
```

```
of << str.in() << endl;
of.close ();


printf ("Running.\n");

mgr->activate ();
orb->run();
poa->destroy (TRUE, TRUE);
delete micocash;

return 0;
}
```

6.3.5.2   The Servant

In most application scenarios, the target would like to know which client invokes its operations. This check cannot be done in the server main function because there is no exact association between the server and its clients. The security information in the server main function is associated with the ORB instance or the server process as a whole.

The POA dispatches requests from the clients to *servants*, so within the servant (i.e., the actual implementation of the operations in the server application), there is a clear secure association with the client that invoked the operation. Consequently, the servant is the right place to ask the ORB about the client's security attributes; in the servant, this information is specific to the thread of execution. There might be more than one servant thread, with every thread associated with a different client and a different set of security attributes. These thread-specific `Credentials` have to be accessed through `Current` and not through the process-specific `SecurityManager` object, which was used in the server part of the target application.

This section will illustrate how the client's security attributes can be accessed from within the servant. The first code fragment, which implements the operations `bal` and `deposit`, remains unchanged from the nonsecured version:

```
class Account_impl : virtual public POA_Account
{
public:
  Account_impl ();
```

```
    void deposit (CORBA::ULong);
    void withdraw (CORBA::ULong);
    CORBA::Long balance ();

private:
  CORBA::Long bal;
};

Account_impl::Account_impl ()
{
  bal = 0;
}

void
Account_impl::deposit (CORBA::ULong amount)
{
  bal += amount;
}
```

The security-related functionality that retrieves the security attributes of the invoking client is part of the withdraw implementation:

```
void
Account_impl::withdraw (CORBA::ULong amount)
{
```

First of all, we need to get the SecurityCurrent object and narrow it to the correct data type:

```
CORBA::Object_var securitycurrent;
SecurityLevel2::Current_var seccur;
securitycurrent =
  orb-> resolve_initial_references
     ("SecurityCurrent");
  assert (!CORBA::is_nil (securitycurrent));

seccur = SecurityLevel2::Current::
  _narrow(securitycurrent);
assert (!CORBA::is_nil (seccur));
```

In the servant, we are interested in the attributes of the client that invoked the servant. To get that information, we need to obtain the `Received-Credentials` object, which was transferred from client to target during the secure association establishment:

```
SecurityLevel2::ReceivedCredentials_var rc =
    seccur->received_credentials();
```

Now it is possible to access the security attributes from the credentials as previously described. Again, we create an instance of a security attribute list and fill it with the requested attribute:

```
Security::ExtensibleFamily fam;
fam.family_definer = 0;
fam.family = 1;
Security::AttributeType at;
at.attribute_family = fam;
at.attribute_type = Security::AccessId;
Security::AttributeTypeList atl;
atl.length(1);
atl[0]=at;
```

Then we call `get_attributes` on the `ReceivedCredentials` object to retrieve the specified attribute:

```
Security::AttributeList_var al =
rc->get_attributes( atl );
```

Finally, the content of the retrieved attribute is printed out on the console:

```
cout << "Received  credentials"
    << (*al).length() << " attributes\n";
for ( int ctr = 0; ctr (*al).length(); ctr++) {
  cout
    << (*al)[ctr].attribute_type.
       attribute_family.family
    << " "
    << (*al)[ctr].attribute_type.attribute_type
```

```
        << " "
        << &(*al)[ctr].value[0] < " "
        << &(*al)[ctr].defining_authority[0]
        << endl;
  }
```

From here, the rest of the servant code remains unchanged. It implements the `balance` operation and the `Bank` interface with its `create` operation:

```
bal -= amount;
}

CORBA::Long
Account_impl::balance ()
{
  return bal;
}


class Bank_impl : virtual public POA_Bank
{
public:
  Account_ptr create ();
};

Account_ptr
Bank_impl::create ()
{

  Account_impl * ai = new Account_impl;

  Account_ptr aref = ai->_this ();
  assert (!CORBA::is_nil (aref));


  return aref;
}
```

6.3.6   The Client

This section examines the security enhancements in the client source code. In many respects, the client-side use of level 2 interfaces is quite similar to the server part. In particular, SecurityManager is used throughout the client code.

The client first resolves the initial object, the SecurityManager. Then it calls the PrincipalAuthenticator to create its OwnCredentials, using the security information from a X.509 certificate, and prints the credentials content out on the console. In the next step, it binds to a server and prints several attributes from the TargetCredentials (i.e., from the remote peer of the secure association) before it invokes operations on the target. We will now discuss the security-related extensions of the client code in more detail:

```
#include "account.h"

#ifdef HAVE_UNISTD_H
#include <unistd.h>
#endif

CORBA::ORB_var orb;
CORBA::Object_var securitymanager;
SecurityLevel2::SecurityManager_var secman;

int
main (int argc, char *argv[]) {
  CORBA::ORB_var orb =
      CORBA::ORB_init (argc, argv, "mico-local-orb");
```

As usual, we first resolve the reference to the SecurityManager object:

```
securitymanager =
  orb-> resolve_initial_references
    ("SecurityManager");
secman = SecurityLevel2::SecurityManager::
  _narrow(securitymanager);
assert (!CORBA::is_nil (secman));
```

Next, the authentication method has to be specified. Again, this is done exactly as in the previously described code examples for the target:

```
Security::AuthenticationMethod our_method =
  (Security::AuthenticationMethod)SecurityLevel2::
KeyCertCAPass;
```

Once the method has been set, the client has to create a variable for the authentication data and feed information about the client's security information into it. In MICOSec, the file names of X.509 certificate and key files have to be provided as follows:

```
Security::SSLKeyCertCAPass *method_struct;
method_struct = new Security::SSLKeyCertCAPass;
CORBA::Any* any_struct =
  secman -> get_method_data(our_method);
*any_struct >>= *method_struct;
method_struct -> key = "ClientKey.pem";
method_struct -> cert = "ClientCert.pem";
method_struct -> CAfile = "";
method_struct -> CAdir = "";
method_struct -> pass = "";

CORBA::Any* out_any_struct;
out_any_struct = new CORBA::Any;
*out_any_struct <<= *method_struct;
```

The client then calls the authenticate operation on the `Principal-Authenticator`, which creates its `OwnCredentials` object and adds it to the `OwnCredentials` list in `SecurityManager`:

```
SecurityLevel2::PrincipalAuthenticator_ptr pa =
  secman -> principal_authenticator();
const char* mechanism = "";
const char* security_name = "ssl";
Security::AttributeList privileges;
SecurityLevel2::Credentials_ptr creds;
CORBA::Any* continuation_data;
CORBA::Any* auth_specific_data;

try
```

```
    {
        pa -> authenticate( our_method, "IDEA-CBC-SHA",
            security_name, *out_any_struct,privileges,
            creds,continuation_data,auth_specific_data);
    }
    catch (...)
    {
        cout << "authentication failed" <<endl;
        delete out_any_struct;
        return 0;
    }


    delete out_any_struct;
```

Now, the client obtains its OwnCredentials, which have been put in as the first element in the SecurityManager's own_credentials attribute list. Before the attribute content can be obtained, a security attribute type list has to be instantiated and filled with the requested security attribute:

```
SecurityLevel2::Credentials_ptr own_cred;
own_cred = (*(secman -> own_credentials()))[0];

Security::ExtensibleFamily fam1;
fam1.family_definer = 0;
fam1.family = 1;
Security::AttributeType at1;
at1.attribute_family = fam1;
at1.attribute_type = Security::AccessId;
Security::AttributeTypeList atl1;
atl1.length(1);
atl1[0]=at1;
```

We can now get the security attributes from the Credentials and print them out on the console:

```
Security::AttributeList_var al1 =
    own_cred->get_attributes(atl1);
```

```
cout << "Own credentials"
    << (*all).length()
    << " attributes\n";

for ( int ctr = 0; ctr (*all).length(); ctr++) {
  cout
    << "family = "
    << (*all)[ctr].attribute_type.
       attribute_family.family
    << " "
    << "type = "
    << (*all)[ctr].attribute_type.attribute_type
    << " "
    << &(*all)[ctr].value[0]
    << " "
    << &(*all)[ctr].defining_authority[0]
    << endl;
}
```

This part just reads the stringified IOR from the file `Bank.ref`, converts it back to an object, and binds to the `Bank` target application:

```
char pwd[256], uri[300];
sprintf (uri, "file://%s/Bank.ref",
  getcwd(pwd, 256));



CORBA::Object_var obj = orb->string_to_object (uri);
Bank_var bank = Bank::_narrow (obj);
assert (!CORBA::is_nil (bank));
```

Now that a connection to the bank has been established, the client can obtain the target's security attributes in a similar way as described for the `OwnCredentials` above. These `TargetCredentials` contain the security attributes of the target, but not directly. Normally, they are just a reference to an object with security mechanism-specific content. Again, we first create a security attribute type list with the requested attribute:

```
Security::ExtensibleFamily fam2;
fam2.family_definer = 0;
fam2.family = 1;
Security::AttributeType at;
at.attribute_family = fam2;
at.attribute_type = Security::AccessId;
Security::AttributeTypeList atl;
atl.length(2);
atl[0]=at;
Security::AttributeType at2;
fam2.family = 11;
at2.attribute_family = fam2;
at2.attribute_type = Security::PeerAddress;
atl[1]=at2;
```

So far, the client-side modifications to the code look very similar to the ones for the target. Indeed, the process for getting the initial object Security-Manager and the OwnCredentials is identical to the target.

However, there is one notable difference to the target side: A servant has exactly one secure association with the client on whose behalf the servant is called. Therefore, it is clear which attributes the application wants to obtain in the servant.

On the client side, the situation is different. A client might call operations on several different targets, and several secure associations with different servers and hosts might exist. Consequently, it is necessary on the client side to tell the security service for which secure association the attributes should be obtained. This is done by specifying the secure association through the corresponding IOR of the target:

```
SecurityLevel2::TargetCredentials_var tc =
    secman->get_target_credentials(bank);
```

The parameter bank is the IOR of the target server, and the operation get_target_credentials on the SecurityManager obtains the TargetCredentials. We can get the target's security attributes from it in the usual fashion and print them out on the console:

```
Security::AttributeList_var al =
  tc -> get_attributes(atl);

cout << "Server credentials "
     << (*al).length() << " attributes\n";

for ( int ctr = 0; ctr (*al).length(); ctr++) {
  cout
     << "family = "
     << (*al)[ctr].attribute_type.
        attribute_family.family
     << " "
     << "type = "
     << (*al)[ctr].attribute_type.attribute_type
     << " "
     << &(*al)[ctr].value[0]
     << " "
     << &(*al)[ctr].defining_authority[0]
     << endl;
}
```

The rest of the client code is unchanged. It creates an account object on the target side and invokes a few operations, including withdraw, which trigger the security code on the target side:

```
Account_var account = bank->create ();
if (CORBA::is_nil (account)) {
    printf ("oops: account is nil\n");
    exit (1);
}


cout  << "deposit - 700\n";
account->deposit (700);
cout  << "withdraw - 450\n";
account->withdraw (450);
printf ("Balance is %ld.\n", account->balance ());
return 0;
}
```

6.4 Object Domain Mapper for Access Control and Audit

6.4.1 Introduction

MICOSec's access control and audit components are implemented on the middleware layer (i.e., on top of the SSL-based principal authentication, security context establishment, and message protection functionality described so far).

Whenever a request arrives at the target, MICOSec checks its access control policy to see if the access is allowed. If it is allowed, then the invocation is passed on to the servant; otherwise, it is rejected (access control will be described in Section 6.5). The audit policy is also checked for each invocation to see if the action conveyed in the request should be logged (as described in Section 6.6). This implicitly assumes that it is easy to find the correct security policy associated with the target that is being invoked. However, due to the unpredictable and transient nature of the information in the request header, this turns out to be a difficult task.

This section describes how MICOSec solves this problem by introducing an object domain mapper (ODM), which helps find the access control or audit policy associated with the invoked target. ODM is a precondition for the MICOSec level 2 access control and audit functionality.

On the application layer (i.e., for security-aware applications), access control and audit are done inside the application, and so the policy is already associated with the correct target.

At the ORB layer (i.e., for security-unaware applications), the association has to be set up using the information available from the request header and security service. The standard CORBA security access control and audit models (described in Chapter 3) use object interface types to describe targets in their security policies. This approach is simple, but it has a number of disadvantages: First, the same access policy applies to all objects of the same type, which is insufficient for most real-world applications. Second, it is not always possible to determine the most derived interface of an object, because any operation behind an interface could be implemented by its parent class.

The solution described here is based on an extended version of the ODM described as part of the Security Domain Membership Management Service (SDMM) [1]. It uses unchanging domain names to describe target objects persistently inside the security policies. Whenever an invocation arrives, the information about the target is extracted from the request header (plus identity information from the security service) and mapped onto the domain name. This way, the correct policy for the invoked target can be looked up based on this unchanging ODM domain name.

The main design requirements for this ODM are:

- *Object life cycle independence:* The administration of the mapping configuration needs to be independent of the object life cycle. For example, it should be possible to define the mapping before the object has been launched.

- *Performance:* The mapping needs to be fast because it has to be done for each incoming request.

- *Scalability, ease of administration:* Grouping objects that have the same policy into domains allows easy administration of policies in large-scale applications.

- *Flexibility of use:* The mapping should work for different types of CORBA applications (e.g., for different activation modes) and without any modifications to the application code.

- *Independence from the application:* It should be possible to do the mapping without activating the servant, so that the policy can be enforced before the servant is even activated.

- *Ease of integration into the architecture:* Changes to the ORB and security services should be kept at a minimum, to allow for portability and replaceability.

- *Trust by the target:* The target has to trust all information used as part of the mapping process, so the ODM process needs to be kept inside the target trust domain.

6.4.2 Mapping Information

As a first step, the ODM has to find the right domain name for each incoming invocation. The information available on the ORB layer is often called the *surrogate* for the associated target object. In this section, we examine the different options for surrogates: the object reference, security attributes, and the POA name.

6.4.2.1 Request Header

The IOR is generated by the POA when the servant is registered with the POA and contains the information necessary for the client to be able to invoke the target. The following information inside the IOR is used by the client to generate the GIOP request header for the invocation:

- *Repository ID:* The repository ID is a simple string that describes the interface type of the object in a standardized format. This type is the most derived interface (MDI) of the object at the time of its instantiation. However, this type may not reflect the real object type at the time of the invocation, because as a consequence of changes in the repository, it is possible that another object (with a derived interface) contains the implementation of the invoked operation instead.

- *Endpoint:* The IOR also contains transport-specific endpoint information to locate the server. In the case of IIOP, this endpoint is a TCP socket `<hostname, port>` pair to which the transport should open a TCP connection. Note that this endpoint does not necessarily connect directly to the target; it could instead point to an activation daemon (i.e., an implementation repository) that returns a *location forward.* This location forward points to the actual target server. Alternatively, the IOR could point to a firewall, or it could contain information about several endpoints. The format of such endpoints is standardized in the CORBA specification and depends on the underlying transport (e.g., TCP/IP).

- *Object key:* Once the client has established a network connection to the server, it needs to address the correct target object within the server process. Depending on the POA policy, the object key in most cases contains a random identifier of the POA, which is unique within each server, and a random object identifier, which uniquely identifies the servant object within the scope of the specified POA. For scalability reasons, the POA identifier is often chosen at random by the ORB when the POA is created, and, in most cases, the object identifier is also randomly chosen by the POA when the object is activated. The object key format is not standardized by CORBA. This means that only the POA that registered the object will be able to interpret the object identifier correctly.

6.4.2.2 X.509 Identity

The server can also be described by its security attributes. SSL describes the endpoints of a network connection with attributes from X.509 identity certificates (i.e., independently from the actual physical address). This has a number of advantages: The SSL context can traverse transport layer firewalls, and several servers can have the same identity (i.e., load balancing and redundant services can easily be implemented).

6.4.2.3 POA Name

POAs are arranged in a hierarchy, with a so-called *root POA* as a root to a tree of child POAs. Each POA in the tree has operations that allow the insertion or deletion of new POAs in the tree. In this hierarchical structure, each POA has a unique, persistent *POA name*, which can be obtained at run-time.

6.4.3 MICOSec Mapping Definition

The main goal is to define fine-grained persistent mapping rules for security-unaware applications. The security administrator should be able to define the mapping between surrogates (which describe objects or groups of objects) and their associated domains. This should be possible independently from the application object life cycle and without any modifications to the source code. The basic configuration and code example described in this section shows how to set up and use MICOSec's object-to-domain mapping mechanism. The mapping is defined in a configuration file that is read into MICOSec to set up a mapping table when the application is launched. It is also possible at run-time to dynamically create or modify the mapping table. In addition, it is possible to save the content of the mapping table back into a persistent file for future use.

The mapping itself can be defined at several levels of granularity:

- Default;
- X.509 subject;
- POA name;
- Object identifier.

In most real-world applications, the first three levels of granularity are sufficient. In this section, we will discuss how each is defined in MICOSec, while the more unusual ODM on a per-object granularity will be described in Section 6.4.7.

The lowest level of granularity is specified by a `Default` domain, which applies to all objects if no more specific domain mapping rules are defined in the configuration.

The next level of granularity is defined by the `AccessId` security attribute provided by the underlying security mechanism. In the case of MICOSec, the `AccessId` contains the subject of the server's X.509 certificate. Although CORBA security also defines an audit-specific `AuditId`,

MICOSec only uses the `AccessId` to express principals for both access control and audit. This makes sense because SSL normally associates a single X.509 identity with each principal, which means that both `AccessId` and `AuditId` would contain the same value anyway.

By using the certificate-based `AccessId`, the server can be described logically inside access control and audit policies (i.e., independently from the hostname of the physical server location). This allows the elegant implementation of load balancing and redundancy because several physically separate servers can have the same logical identifier and will, therefore, be mapped onto the same domain name. Analogously, several CORBA applications with separate X.509 identities can run on a single server because the certificate to use can be specified at the command line when the application is launched, or in the application source code itself.

The granularity of `AccessId` depends on the underlying security mechanism. In the case of MICOSec (and many other implementations), SSL is used to authenticate network endpoints. As a result, all objects behind a TCP endpoint will have the same SSL context (i.e., the same security attributes) and will consequently be mapped onto the same domain name.

The persistent configuration of this mapping is easy. The X.509 certificate content is already known before the application is even launched, so it can be stored in a persistent configuration file without any difficulties.

At run-time, the X.509 certificate content can easily be read out of the security mechanism (through the `Credentials` object) and be used to search for the corresponding domain name in the ODM table.

Now that the endpoint (i.e., the ORB) has been logically defined by its security attributes, the actual target object needs to be specified at a finer granularity. The only information available from the IOR is the object key, which contains a reference to the POA that generated the object, and the object identifier, which describes the exact object within the realm of the POA. Unfortunately, the POA identifier is only unique within the scope of its underlying ORB, and the object identifier is chosen randomly by the POA to allow for good scalability. Only the ORB can interpret the POA identifier and locate the correct POA. In the same way, only the POA can interpret the object identifier and pass the request up to the object. Because of this, neither the POA identifier nor the object identifier can be used for persistent mapping.

To solve this problem, MICOSec instead exploits the fact that POAs can be arranged in a hierarchy, with a so-called root POA as a root to a tree of child POAs. As part of this hierarchical structure, each POA has a unique, persistent name (which is different from the POA identifier), that can be obtained at run-time. By using this POA name, MICOSec can support a transparent, persistent mapping on the granularity of the POA. Section 6.4.3.1 illustrates how the ODM is set up.

6.4.3.1 MICOSec Mapping Definition Example

Every application object resides on top of a leaf of the POA hierarchy (which can also consist of only a single node), with the so-called root POA being the root of the tree. The root POA has a well-defined default policy that describes aspects related to object creation. It is possible to get a reference to this root POA through its initial reference as follows:

```
CORBA::Object_var poaobj =
    orb -> resolve_initial_references ("RootPOA");
PortableServer::POA_var poa =
    PortableServer::POA::_narrow (poaobj);
```

This POA automatically has the persistent name RootPOA, which is not related to the randomly chosen POA identifier put in the object key. In this example, the reference to the root POA is called poa. Starting from the root POA, a hierarchy of child POAs can be created. This example creates a child POA object called "MyPOA" under the "RootPOA" (mgr is the POA manager and pl is the POA policy):

```
PortableServer::POA_var mypoa =
    poa-> create_POA ("MyPOA", mgr, pl);
```

The child POA "MyPOA" has the complete hierarchical name "/Root-POA/MyPOA" that describes the POA uniquely and persistently. The crucial point is that it does not matter if the POA itself is persistent or transient—this hierarchical name is independent from the identifier put into the IOR's object key. This allows administrators to define POAs independently from the application life cycle.

To make full use of this POA hierarchy for the ODM, it needs to reflect the required domain hierarchy. In other words, objects that should belong to the same security policy domain should be grouped into the scope of the same POA. The following MICOSec configuration file shows how the ODM is set up for the per-POA level of granularity:

```
# POA-Mapping                              Domain
# Keys: AccessId, POA


# Default, always applies if
# no mapping was defined


[<Security Attribute>]/                    /ObjectSecurity


# The different POAs and their Domains


[<Security Attribute>]/RootPOA/            /ObjectSecurity/Domain
[<Security Attribute>]/RootPOA/MyPOA/       /ObjectSecurity/Domain1
[<Security Attribute>]/RootPOA/MyPOA2/      /ObjectSecurity/Domain2
[<Security Attribute>]/RootPOA/AccountPOA/  /ObjectSecurity/Accounts
```

Persistent POA Mapping

Such an ODM configuration file has to be created manually by the administrator and made accessible to MICOSec at the time the application is launched. The example described below also shows how the ODM configuration can be modified at run-time and stored back into a file. In Section 6.4.8, we will show how the ODM configuration can be created or modified at run-time from the application layer.

6.4.4 Mapping Process

Section 6.4.3 showed how the mapping from POA names to domain names is configured in MICOSec. This section examines where in the architecture the actual mapping is carried out and how it is achieved.

In most cases, the right place to do the mapping is where the security policy has to be enforced. You will learn later in this chapter how MICOSec's access control and audit policies make use of ODM domain names. If application layer security is used, then the domain name needs to be determined by the servant. On the ORB layer, the ODM can be integrated into the part of the security subsystem that is called whenever an invocation occurs (i.e., inside the interceptor implementation). Regardless of whether the ODM is used from the interceptor or the servant, the domain needs to be determined within the context of the incoming invocation. The ORB makes this execution context–specific information available to interceptors and servants though the `Current` object. For example, `SecurityCur-rent` provides information about the client that invoked the operation.

Moreover, the full POA name can simply be obtained from `POA-Current` and mapped directly onto the domain name specified in the configuration file. This is much more elegant and useful than obtaining the target of an invocation from `Current`, determining the ORB-specific (random) POA reference from the request header, and trying to map it (somehow) onto a static domain name.

There is an interesting conceptual relationship between the ODM described here and the CORBA naming service. Analogously to the domain name service on the Internet, the CORBA naming service resembles the so-called *forward resolving*, whereas the ODM resembles *reverse resolving*. The naming service maps a name to an IOR, while the ODM maps—at least conceptually—an IOR back to a name. However, in practice these two names have quite different semantics, which complicates the conversion of domain names into IORs and vice versa and, thus, makes it necessary to use MICOSec's ODM feature.

6.4.5 ODM Interfaces

The ODM includes interfaces for objects that represent domains and for the actual domain mapper. The format of domains is defined in `Domain-Manager.idl` (as specified in [1]):

```
module SecurityDomain{
  typedef string Istring;

struct NameComponent{
    Istring id;
    Istring kind;
};

  typedef sequence <NameComponent> Name;
  typedef sequence <Name> NameList;
  typedef unsigned short PolicyCombinator;
  interface NameIterator;
```

A domain is defined by a sequence of names, with each containing two strings: The identifier string contains the component in the domain path, and the type string can contain an unspecified description of the component.

The interfaces for the ODM, the ODM manager, and the actual ODM mapper are defined in odm.idl (also specified as part of [1]). The first interface Manager contains operations to set domain names, get domain names, remove domain names, set the parent ODM key, set the default parent ODM, and set the default name key:

```
interface Manager {
  typedef unsigned short ODMGranularity ;

void set_domain_name_key (
  in ODMGranularity granularity,
  in Security::Opaque key,
  in SecurityDomain::NameList domainNameList
);

SecurityDomain::NameList get_domain_names (
  in Security::Opaque key
);

void remove_domain_names (
  in Security::Opaque key
);

void set_parent_odm_key (
  in Security::Opaque key,
  in ObjectDomainMapping::Manager odm
);

void set_default_parent_odm (
  in ObjectDomainMapping::Manager odm
);

void set_default_name_key (
  in SecurityDomain::NameList domainNameList
);
};
```

The main task of the following ODM factory is to create the ODM manager. In addition, the MICOSec-specific interface was extended by operations to load and save ODM configuration files:

```
interface Factory {
  ObjectDomainMapping::Manager create ();

  //extension
  boolean loadConfigFile(in string filename);
  boolean saveConfigFile(in string filename);

};

interface ODM {
  ObjectDomainMapping::Factory create ();

  };

};

#endif
```

6.4.6   Static Per-POA Granularity

This section illustrates how the mapping process is used. The use is the same in the servant (i.e., for security-aware applications) and interceptor (i.e., for security-unaware applications).

First, references to the underlying ORB and the SecurityManager are created. Note that this does not create the ORB but rather establishes a pointer to the existing underlying ORB:

```
CORBA::ORB_var orb;
CORBA::Object_var securitymanager;
SecurityLevel2::SecurityManager_var secman;

orb = CORBA::ORB_instance ("mico-local-orb", FALSE);

securitymanager =
  orb-> resolve_initial_references
    ("SecurityManager");
assert (!CORBA::is_nil (securitymanager));
secman = SecurityLevel2::SecurityManager::
  _narrow(securitymanager);
  assert (!CORBA::is_nil (secman));
```

Next, the `AccessID` of the target has to be determined, which is part of the `own_credentials` object of the target. This is done in exactly the same way as described in Section 6.3. First, a reference to the `OwnCredentials` has to be created:

```
SecurityLevel2::Credentials_ptr own_cred;
own_cred = (*(secman -> own_credentials())) [0];
```

Then an attribute type family `fam1` is created and set to specify which attributes should be obtained (i.e., the `AccessID`):

```
Security::ExtensibleFamily fam1;
fam1.family_definer = 0;
fam1.family = 1;
Security::AttributeType at1;
at1.attribute_family = fam1;
at1.attribute_type = Security::AccessId;
Security::AttributeTypeList atl1;
Security::AttributeTypeList atl1;
atl1.length(1);
atl1[0]=at1;
```

Now the `AccessID` security attribute can be obtained from the `Own-Credentials` object:

```
Security::AttributeList_var al1 =
   own_cred-> get_attributes(atl1);
```

The obtained identity is now stored in a string `key1` in a format that contains brackets before and after the attribute content:

```
string key1 = "[";
   for ( int ctr = 0; ctr < (*al1).length(); ctr++)
      {
         key1 += (char *)(&(*al1)[ctr].value[0]);
      }
   key1 += "] ";
```

Next, the POA of the current execution context is determined by calling the operation `get_POA` on the reference to the `POACurrent`:

```
CORBA::Object_var poao =
  orb-> resolve_initial_references
      ("POACurrent");
PortableServer::Current_var cpoa =
  PortableServer::Current::_narrow(poao);
PortableServer::POA_ptr poa = cpoa->get_POA();
```

Next, a few variables are created that are needed to determine the full name of the POA (i.e., the full path through the POA hierarchy by iterating through the POA hierarchy). This iteration process starts at the actual POA leaf that the object resides on and appends the name of each underlying POA to a string `tstr` until the root POA is reached. At the end of this process, `key1` contains the full POA name in the same format as specified in the ODM configuration table:

```
PortableServer::POA_ptr np = poa;
string key2;
string tstr;


while (np != NULL) {
   tstr = np->the_name();
   if (key2.length()  0)
      tstr += '/';
   tstr += key2;
   key2 = tstr;
   np = np->the_parent();
}
cout << "POA=" << key2 << endl;

key1 += key2;
```

For the actual mapping, a pointer to the `ODMManager` is required. If an `ODMManager` is available, then the string `key1` that contains the full POA name is copied into a variable of the type `Opaque`. Once this is done, the operation `get_domain_names` can be invoked on the `ODMManager` to obtain the domain name associated with the POA name:

```
ObjectDomainMapping::Manager_ptr dmanager =
    poa->get_ODM();
 if (dmanager) {

    Security::Opaque okey;
    int len = key1.length();
    okey.length(len);
    for (int i = 0; i < len; i++)
      okey[i] = key1[i];

    SecurityDomain::NameList * list =
       dmanager-> get_domain_names(okey);
```

The operation get_domain_names returns a list of the domain names of which the target object is a member. As a last step, the domain names can be printed out on the console like this:

```
cout  << "Domain=";
for (int i = 0; i < (*list).length(); i++) {
    SecurityDomain::Name nm = (*list)[i];
    for (int j = 0; j nm.length(); j++) {
        SecurityDomain::NameComponent nc = nm[j];
        cout << "/" << nc.id;
    }
    cout << endl;
}
```

The purpose of this code example was to describe the basic ODM functionality. MICOSec does not come with an explicit ODM example, because Sections 6.5 and 6.6 already demonstrate the use of the MICOSec ODM in the context of level 2 access control and audit.

6.4.7   Per-Object Granularity

In some cases, per-object granularity ODM is required to express specific security policies for individual objects. However, persistent mapping at the level of object instances cannot be achieved elegantly, because CORBA does not assign persistent names to objects. Therefore, it is preferable for most real-world applications to instead match the POA hierarchy with the required granularity.

If per-object granularity is needed, then substantial modifications to the application source code are necessary, and the `IDAssignmentPolicy` of the `POA` needs to support the `USER_ID` policy, which (as opposed to the `SYSTEM_ID` policy) allows applications to set their own object identifier in the IOR themselves.

To configure per-object granularity ODM, the object identifier has to be appended to the `<Security Attributes, POA name>` pair in the configuration file, and a domain name needs to be specified for it. If the mapping should be persistent, then the application-specific identifier needs to match the identifier that has been put into the configuration file. During the mapping process, the object identifier needs to be added to the string that is supplied to the ODM. If no match is found, then the search is automatically repeated without the object identifier (i.e., on a per-POA granularity).

6.4.8 Dynamic Configuration

The per-POA granularity example in Section 6.4.6 defined a persistent mapping between object and security domains in a configuration file. In some cases, it may be necessary during the application lifetime to dynamically create or modify the mapping or to add application-specific object identifiers into the mapping configuration file at run-time. This section illustrates how this is done inside the servant implementation.

First, an ODM factory for the ODM needs to be created by invoking `create` on the narrowed initial reference to `ODM`:

```
CORBA::Object_var objodm =
  orb-> resolve_initial_references ("ODM");
ObjectDomainMapping::ODM_var odm =
  ObjectDomainMapping::ODM::_narrow(objodm);
ObjectDomainMapping::Factory_var factory =
  odm-> create();
```

Next, a child POA called "MyPOA" is created, which registers the generated ODM factory:

```
PortableServer::POA_var mypoa =
  poa-> create_POA ("MyPOA", mgr, pl);
mypoa->registerODMFactory(factory);
```

Now the newly created domain manager is associated with the POA. It can be accessed by calling `get_ODM`:

```
ObjectDomainMapping::Manager_ptr dmanager =
    mypoa-> get_ODM();
```

In this example, we would like to query if a domain exists in the configuration file for a defined key. Later, we will show how new keys can be inserted into the configuration. If required, a preexisting configuration file could be read into the ODM by calling the `loadConfigFile` operation on the ODM factory before this is done.

To ask the ODM if a domain name exists for a key, we first have to define the key `okey` that has to be searched for:

```
string key = "[/C=UK/ST=Server
State/L=Cambridge/O=ObjectSecurity Ltd./OU=RD/C
N=ServerTest/Email=server@test] MyPOA";
Security::Opaque okey;
int len = key.length();
okey.length(len);
for (int i = 0; i < len; i++)
        okey[i] = key[i];
```

Once this is done, the list of domain objects can be retrieved and the domain names can be extracted as follows:

```
SecurityDomain::NameList * list =
    dmanager-> get_domain_names(okey);

    for (int i = 0; i < (*list).length(); i++) {
        SecurityDomain::Name nm = (*list)[i];
        for (int j = 0; j nm.length(); j++) {
            SecurityDomain::NameComponent nc = nm[j];
            cout << j << "    " << nc.id << endl;
        .}
    }
```

It is also possible to define a new mapping in the configuration that involves the following three steps. First, the new key has to be defined and stored in a suitable variable:

```
Security::Opaque okey2;
string key2 = "[/C=UK/ST=Server
State/L=Cambridge/O=ObjectSecurity
Ltd./OU=RD/CN=ServerTest/Email=server@test]
MyPOATR";
len = key2.length();
okey2.length(len);
for (int i = 0; i < len; i++)
        okey2[i] = key2[i];
```

Next, a hierarchical domain name for the key has to be created. In this example, the hierarchy contains only a single layer:

```
SecurityDomain::NameList dl;
dl.length(1);
SecurityDomain::NameComponent nc;
nc.id = CORBA::string_dup("New Domain");
nc.kind = CORBA::string_dup("");
SecurityDomain::Name nm;
nm.length(1);
nm[0] = nc;
dl[0] = nm;
```

Finally, the entry has to be inserted into the mapping table by calling set_domain_name_key. This operation allows the modification or complete redefinition of the mapping table at run-time:

```
dmanager2->set_domain_name_key
        ((CORBA::UShort)2, okey2, dl);
```

If needed, the new ODM configuration can also be stored in a file:

```
factory->saveConfigFile("newconfig.cnf");
```

6.4.9 Modifications to the CORBA Specification

One of the goals of the ODM architecture was to keep modifications to the existing CORBA specification as minimal as possible to fulfill CORBA's portability and replaceability requirements.

Only a few modifications to the MICO ORB were necessary; in particular, two functions to the POA to allow the registration of an ODM factory with a POA and to query the ODM:

```
void registerODMFactory
   (in ::ObjectDomainMapping::Factory fry);

ObjectDomainMapping::Manager_ptr
   MICOPOA::POA_impl::get_ODM()
```

Also, an initial reference "ODM" had to be added, so that a reference to the ODM can be obtained. This reference is necessary to allow access to the ODM from within interceptors and servants. Without this, features like dynamic configuration could not be implemented.

6.5   Access Control

Access control in CORBA is based on access policies that are associated with the targets to be protected. These policies are evaluated and enforced whenever an invocation arrives. As part of the process, the rights required to access the target are compared to the rights granted to the calling principal. If the principal's rights match the target's required rights, then access is allowed. Otherwise, the invocation is rejected (see Section 3.3.6 for more details).

CORBA security access control mainly involves the following objects:

- AccessDecision carries out the evaluation of the access policy.
- RequiredRights states the target access policy.
- DomainAccessPolicy contains the rights associated with calling principals.
- DomainManager is responsible for managing this (and other) policies for a domain.

Whenever a request arrives at the target ORB, the security service queries the DomainAccessPolicy object for the rights granted to the CORBA principal associated with the request's mechanism-specific security context information. The DomainAccessPolicy for each domain is managed by that domain's DomainManager. The security service also obtains the rights required to access the invoked target object type and operation from the RequiredRights object. Based on this information, the AccessDecision object allows access only if the principal's access rights from `DomainAccessPolicy` are sufficient to meet the target's `RequiredRights`.

Therefore, an early version of MICOSec implemented domain-based access control according to the CORBA security specification (see 3.3.7). But the practical evaluation identified a number of issues: First, it was unclear how the access control decision object could find the required rights associated with a target object, as neither the IOR nor the interface type is useable in practice (see 6.4.1). As a result, the domain name (which we get from the ODM) had to be used for this purpose. In addition, it was deemed not very intuitive to map target objects to domains, and then use these domains to map principals to their granted rights. Target objects, their domains, and their associated security policies are one logical unit. Also, from a more technical perspective, the domain-based mapping of principals to granted rights unnecessarily complicates the integration of directory services that store user's rights.

In MICOSec, these issues were solved by changing the access control (nonconformant to the current specification) in such a way that the mapping of the principal's attributes to its granted rights is not domain based. Instead, a principal simply has a set of granted rights to invoke operations on all objects. On the target side, objects are mapped to domains that are associated with the security policy to apply and with the required rights. Note that these issues are not limited to access control; the audit service also has to find the object's audit policy. We will see later (in section 6.6) that the problems can be solved there in a similar way.

The standard `AccessDecision` object deals with access rights on a very coarse granularity: `get`, `set`, `use`, and `manage` are the only specified standard rights used to describe access to particular operations. It is also possible to define additional access rights to express more specific policies. Figure 6.4 illustrates how the main components of the access control model are related.

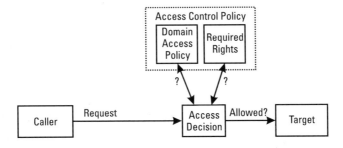

Figure 6.4 ORB layer access control.

In MICOSec's access control policy, the target operations to which the policy should apply are expressed by the target's operation name and domain name. Domain names are supplied by the ODM, which maps the target's X.509 identity and POA name in the hierarchy (and maybe also its object identifier) into its domain name. MICOSec allows for any letter of the alphabet to be used as a right, with the letters "g", "s", "u", and "m" representing the standardized rights `get`, `set`, `use`, and `manage` defined in the CORBA security services specification.

The calling principals are expressed by their X.509 identities and are accessible through the `AccessID` security attribute. It is also possible to cluster users into groups based on the organizational unit (OU) attribute of the X.509 certificate. These groups are represented in CORBA security by the `PrimaryGroupID` security attribute.

The example described in this section shows how the different operations are used to set up an ORB layer access policy from within the server.

6.5.1 Interfaces

The CORBA security services specification defines a number of types related to access control rights (in the file `Security.idl`), in particular, the definition of a right, a list of rights, and the combinators applicable to rights (union or intersection)[5]:

```
struct Right {
     ExtensibleFamily rights_family;
     string rights_list;
   };

typedef sequence <Right> RightsList;

enum RightsCombinator {
     SecAllRights,
     SecAnyRight
   };
```

The interfaces for `RequiredRights` and `AccessDecision` are described in the file `SecurityLevel2.idl`. The `RequiredRights` interface contains operations to get and set the specific rights required to access a particular target operation:

---

5. In the current CORBA security services specification, this is called `right`, but the new CORBA 2.4 IDL does not support this anymore. Therefore, the name was changed to `rights_list`. Note that this is a MICOSec-specific modification to the specification.

```
interface RequiredRights{
  void get_required_rights(
      in Object obj,
      in CORBA::Identifier operation_name,
      in CORBA::RepositoryId interface_name,
      out Security::RightsList rights,
      out Security::RightsCombinator rights_combinator
  );

  void set_required_rights(
      in CORBA::Identifier operation_name,
      in CORBA::RepositoryId interface_name,
      in Security::RightsList rights,
      in Security::RightsCombinator rights_combinator
  );
};
```

The `AccessDecision` interface is called during the access control enforcement to check if the access is allowed or not, depending on the `Credentials` of the caller, the target object, operation, and interface name:

```
interface AccessDecision { // Locality Constrained

  boolean access_allowed (
      in SecurityLevel2::CredentialsList cred_list,
      in Object target,
      in CORBA::Identifier operation_name,
      in CORBA::Identifier target_interface_name
  );
};
```

The `AccessPolicy` interface is described in `SecurityAdmin.idl`. It is the root interface for the various kinds of invocation access control policies. It contains operations to obtain the rights that have been granted to a specified principal:

```
interface AccessPolicy : CORBA::Policy {

  Security::RightsList get_effective_rights (
      in Security::AttributeList attrib_list,
      in Security::ExtensibleFamily rights_family
  );
```

```
        Security::RightsList get_all_effective_rights(
            in Security::AttributeList attrib_list
        );
    };
```

The interface `DomainAccessPolicy` inherits from `AccessPolicy` and provides discretionary access policy management semantics. It has operations to grant, revoke, replace, and get rights for a particular principal:

```
    interface DomainAccessPolicy : AccessPolicy {
        void grant_rights(
            in Security::SecAttribute priv_attr,
            in Security::DelegationState del_state,
            in Security::RightsList rights
        );

        void revoke_rights(
            in Security::SecAttribute priv_attr,
            in Security::DelegationState del_state,
            in Security::RightsList rights
        );

        void replace_rights (
            in Security::SecAttribute priv_attr,
            in Security::DelegationState del_state,
            in Security::RightsList rights
        );

        Security::RightsList get_rights (
            in Security::SecAttribute priv_attr,
            in Security::DelegationState del_state,
            in Security::ExtensibleFamily rights_family
        );

        Security::RightsList get_all_rights(
            in Security::SecAttribute priv_attr,
            in Security::DelegationState del_state
        );
    };
```

6.5.2 The Bank Example

The access control example consists of the usual bank account application, which contains the `Account` interface (with the usual operations `deposit`, withdraw, and `balance`) and the `Bank` interface (with two operations `create` and `open`). In this example, all level 2 access control functionality is configured from within the server and enforced automatically on the ORB layer. The client applications do not contain any access control functionality; their only purpose is to trigger the access control enforcement on the target side.

The example involves three client principals and two target objects with five operations. Each principal has different access rights for each target operation:

- *Manager:* The manager of the application can create bank accounts. To do that, the manager is granted the `manage` right. The example comes with one certificate for the manager.

- *Owner/Wife:* The owner and his wife are users of the bank account application. They require the `use` and `get` rights to open accounts, deposit/withdraw money, and query the balance. The example comes with certificates for both principals, which should both also belong to the same `PrimaryGroupID "family"`.

The target host is also expressed by a certificate.

On the target host, the ODM groups `Bank` objects into the domain "bank" and `Account` objects into the domain "accounts." In this example, it is pure coincidence that objects are grouped into domains depending on their type. Normally, domains and the types of their member objects are not related.

Table 6.1 shows which rights are granted to the manager and owner/wife principal identities and through which standard CORBA security attributes they can be accessed.

Table 6.2 summarizes the content of the target-side access control policies. It states which access rights should be required, depending on the domain name, interface type, and operation name.

6.5.3 Building and Running the Example

The level 2 access control example is compiled by using the `Makefile` in the MICOSec subdirectory `/demo/security/acl-aware`. This directory also contains the X.509 certificates for the three principals (`manager.pem`, `owner.pem`, and `wife.pem`) and other configuration files.

The target application is started by executing the shell script `rss`. It contains command line arguments to set up MICOSec for the example. The first arguments bootstrap SSL, using the server X.509 certificate, and the

Table 6.1
Granted Rights

Security Attribute	Attribute Value (Identities)	Granted Right
AccessID	Manager	Manage
PrimaryGroupID "family"	Owner and wife	Use

Table 6.2
Required Rights

Type/Domain (Policy Combinator `union`)	Interface	Operation	Required Rights (Rights Combinator `any`)
/Access/Bank	Bank	create	Manage
/Access/Bank	Bank	open	Use, get
/Access/Accounts	Account	deposit	Use, set
/Access/Accounts	Account	withdraw	Use, get
/Access/Accounts	Account	balance	Use

server key. Then the parameters for the MICOSec audit service are provided: the ODM configuration file that contains the audit domain information, the audit filter configuration file, the type of audit channel (plain text file), and the name of the log file. Note that no specific command line arguments for access control are supplied; all of the access control functionality is instead set from within the server to demonstrate the use of the level 2 interfaces:

```
./server
-ORBIIOPAddr ssl:inet:'uname -n':12466
-ORBSSLcert ServerCert.pem
-ORBSSLkey ServerKey.pem
-ORBSSLverify 0
-ODMConfig config.cnf
-AuditConfig audit.cnf
-AuditType file
-AuditArchName server.log
-AccessControl on
-Paranoid yes
```

Server shell script

Once the server is running, the client shell script `rcs` can be executed. It starts three different client applications in turn to demonstrate the access control enforcement for all three principals. First, the `client` application is executed with the X.509 certificate of the manager and the SSL private key. Note that in our simplified example, all principals will use the same SSL private key (`key.pem`)—in a real-world application, clients would, of course, have separate keys to protect them from one another. The manager's `client` application invokes the operations `create`, `deposit`, `withdraw`, and `balance`.

Then the `client2` application is launched twice, once with the X.509 certificate of the owner and once with the X.509 certificate of the wife. Both times, it invokes the operations `open`, `deposit`, `withdraw`, and `balance`.

This is the client shell script:

```
#!/bin/sh

ADDR=ssl:inet:'uname -n':12456

echo "Manager"
./client
-ORBBindAddr $ADDR
-ORBSSLcert manager.pem
-ORBSSLkey key.pem
-ORBSSLverify 0

echo "Owner"
./client2

-ORBBindAddr $ADDR
-ORBSSLcert owner.pem
-ORBSSLkey key.pem
-ORBSSLverify 0

echo "Wife"
./client2
-ORBBindAddr $ADDR
-ORBSSLcert wife.pem
-ORBSSLkey key.pem
-ORBSSLverify 0
```

Client shell script

When executed, both the client and target applications produce console output to illustrate how the access control policy is enforced. On the client side, we get the following output, which shows that the policy was enforced correctly: The manager can create accounts, but not deposit/ withdraw or check the balance; the owner and wife can open accounts, deposit/withdraw, and query the balance. Note that there is no relation between creating and opening accounts because accounts are not persistent in this simplified application scenario. The difference is only symbolic to demonstrate how security policies for different operations can be enforced:

```
Manager
Manager started
Manager: Account OK
Couldn't deposit!
Couldn't withdraw!
Couldn't get balance!
Balance is -1.
Owner
Owner/Wife started
Balance is 250.
Wife
Owner/Wife started
Balance is 250.
```

The target-side output shows the result of each access control policy evaluation. Whether access is granted or not depends on the invoked interface and operation, and matches with the policy described previously:

```
Running.
server: after unmarshal for: create
RepoId=IDL:Bank:1.0
+++ allowed! +++
server: after unmarshal for: deposit
RepoId=IDL:Account:1.0
-- denied! --
server: after unmarshal for: withdraw
RepoId=IDL:Account:1.0
-- denied! --
server: after unmarshal for: balance
```

```
RepoId=IDL:Account:1.0
-- denied! --
server: after unmarshal for: open
RepoId=IDL:Bank:1.0
+++ allowed! +++
server: after unmarshal for: deposit
RepoId=IDL:Account:1.0
+++ allowed! +++
server: after unmarshal for: withdraw
RepoId=IDL:Account:1.0
+++ allowed! +++
server: after unmarshal for: balance
RepoId=IDL:Account:1.0
+++ allowed! +++
server: after unmarshal for: open
RepoId=IDL:Bank:1.0
+++ allowed! +++
server: after unmarshal for: deposit
RepoId=IDL:Account:1.0
+++ allowed! +++
server: after unmarshal for: withdraw
RepoId=IDL:Account:1.0
+++ allowed! +++
server: after unmarshal for: balance
RepoId=IDL:Account:1.0
+++ allowed! +++
```

6.5.4 The Target

The file server.cc contains the Bank servant and server implementation. Access control can be evaluated and enforced on the application layer or on the ORB layer. Application-layer access control involves the evaluation of security attributes (provided either by MICOSec or the application itself), and the enforcement of a specific access policy within the application. Application-layer access control will not be explicitly covered because it only involves using the Current/SecurityManager objects to obtain security attributes, which has already been described in detail in Section 6.3. This example illustrates MICOSec's automatic access control enforcement at the ORB layer and, as a consequence, the servant implementation does not contain any security-relevant code.

The ORB layer access control policy can be configured either from the application layer (i.e., from within the server implementation) or transparently from outside the object system by supplying command-line arguments and configuration files. The server implementation of this example illustrates how the ORB layer access control is configured from the application layer, using the level 2 security interfaces.

6.5.4.1 The Servant

As with other examples in this book, the servant contains the implementation of the Bank factory that creates Account objects, which contain operations to deposit and withdraw money and to query the balance. The first part of the servant implementation does not contain any security-related modifications:

```
#include <fstream.h>
#include "account.h"

class Account_impl : virtual public POA_Account
{
public:
  Account_impl ();

  void deposit (CORBA::ULong);
  void withdraw (CORBA::ULong);
  CORBA::Long balance ();

private:
  CORBA::Long bal;
};

Account_impl::Account_impl ()
{
  bal = 0;
}
void
Account_impl::deposit (CORBA::ULong amount)
{
  bal += amount;
}
```

```
void
Account_impl::withdraw (CORBA::ULong amount)
{
  bal -= amount;
}

CORBA::Long
Account_impl::balance ()
{
  return bal;
}
```

The implementation of the Bank object varies slightly from the examples described so far to make use of MICOSec's ODM feature. To use ODM for domain name-based access control, the factory has to register the created Accounts with a different POA in the POA hierarchy. This POA name is then mapped to a domain name by the ODM. The required domain for newly created account objects can be specified by passing the POA associated with the domain name as a parameter to the Bank when it is instantiated.

```
class Bank_impl : virtual public POA_Bank
{
public:
  Bank_impl (){};
  Bank_impl (PortableServer::POA_ptr);
  Account_ptr create ();
  Account_ptr open ();
private:
  PortableServer::POA_var localpoa;
};
```

The POA reference that has been passed into the Bank during instantiation is now duplicated and made available through the variable localpoa:

```
Bank_impl::Bank_impl (PortableServer::POA_ptr _poa)
{
  localpoa = PortableServer::POA::_duplicate (_poa);
}
```

The create operation creates a new Account object and registers it with the POA associated with the domain. It then returns a pointer for the created account to the caller:

```
Account_ptr
Bank_impl::create ()
{

    Account_impl * ai = new Account_impl;
    PortableServer::ObjectId_var oid =
        localpoa->activate_object (ai);
    CORBA::Object_var ref =
        localpoa->id_to_reference (oid.in());
    Account_ptr aref = Account::_narrow(ref);
    return aref;

}
```

In this example, the bank also provides another operation, open, which has exactly the same functionality as create. It has been added to demonstrate how several different operations on the Bank objects can require different access rights. In our example, create will be treated as a function that allows the manager to symbolically open an account for a client (and thus requires manage rights), whereas open is reserved for the owner of the account to open a new account (and needs use and get rights).

```
Account_ptr
Bank_impl::open ()
{
    Account_impl * ai = new Account_impl;
    PortableServer::ObjectId_var oid =
        localpoa->activate_object (ai);
    CORBA::Object_var ref =
        localpoa->id_to_reference (oid.in());
    Account_ptr aref = Account::_narrow(ref);
    return aref;
}
```

6.5.4.2 The Server

The main task of the example server is to set up MICOSec's access control functionality. This includes setting up the POA hierarchy to express the required domains. Then the object domain mapping table is configured based on the POA hierarchy, and a corresponding domain manager hierarchy is created. The domain manager at each node holds the policies applicable to the domain, in this case, the access policy. This access policy can be set with the required rights for each domain. Finally, access rights are granted to the principals by adding them to the associated SecurityManager.

We will now examine the code in more detail. As a first step, the server code initializes the ORB and gets a reference to the RootPOA. The Root-POA will not contain any objects; it just serves as a root to a tree that contains the POAs associated with the domains:

```
int
main (int argc, char *argv[])
{
  CORBA::ORB_var orb = CORBA::ORB_init (argc, argv);
  CORBA::Object_var poaobj =
     orb->resolve_initial_references ("RootPOA");
  PortableServer::POA_var poa =
     PortableServer::POA::_narrow (poaobj);
  PortableServer::POAManager_var mgr =
     poa->the_POAManager();
```

Then it gets initial reference to the ODM and narrows it to the pointer odm. This reference is used to create an ODM factory:

```
  CORBA::Object_var objodm =
     orb->resolve_initial_references ("ODM");
  ObjectDomainMapping::ODM_var
     odm = ObjectDomainMapping::ODM::_narrow(objodm);
  ObjectDomainMapping::Factory_var
     factory = odm->create();
```

The example now loads the configuration file into the ODM factory. If the value NULL is supplied, then the file name from the command line arguments is used. In this example, the configuration file only contains domain mappings for security auditing; all access domain mappings are commented out. Instead, the ODM for access control will be set manually, which illustrates better how access policy mappings can be set from within the server:

```
CORBA::Boolean res = factory->loadConfigFile(NULL);
```

6.5.4.3 POA Hierarchy

The first step involves setting up a POA hierarchy that reflects the domain names. For this purpose, two other POAs—BankPOA and AccountPOA—are created with (empty) default policies. These POAs will be used to describe the domains for Bank and Account objects (the RootPOA will not contain any objects). The BankPOA contains the Bank objects, the factory for Account objects. The AccountPOA contains the Account objects produced by the Bank object:

```
CORBA::PolicyList pl;
pl.length(0);

PortableServer::POA_var bankpoa =
    poa->create_POA ("BankPOA", mgr, pl);
PortableServer::POA_var accountpoa =
    poa->create_POA ("AccountPOA", mgr, pl);
```

To make use of the POA hierarchy for the ODM, all three POAs now have to be registered with the ODM factory:

```
poa->registerODMFactory(factory);
bankpoa->registerODMFactory(factory);
accountpoa->registerODMFactory(factory);
```

6.5.4.4 Object Domain Mapping

The second step configures the object domain mapping. In this example, the ODM is set manually from within the server. First, we get a pointer to the ODM manager, which is used to configure the individual mappings from the X.509 identity of the server host and the full POA names to domain names:

```
ObjectDomainMapping::Manager_ptr dmanager1 =
    poa->get_ODM();
```

Default

We start by configuring the default mapping to the domain `Access`. All target objects that are not explicitly in another domain are automatically put into the `Access` domain. This involves a number of steps: First, the key has to be defined as a string and copied to an `Opaque` data type:

```
Security::Opaque okey2;
string key2 = "[/C=UK/ST=Server
    State/L=Cambridge/O=ObjectSecurity Ltd.
    /OU=RD/CN=Server Test/Email=server@test]";

int len = key2.length();
okey2.length(len);
for (int i = 0; i len; i++)
    okey2[i] = key2[i];
```

Then, a security domain list with only a single list item has to be defined and populated with a name component.

```
SecurityDomain::NameList dl;
dl.length(1);
```

The `NameComponent` is set to have the name "Access" and the type "Access." Note that the type (`kind`) of the `NameComponent` must be set to the function the security domain is used for. In this case, it is set to the type "Access" because it is used as an access control domain.

```
SecurityDomain::NameComponent nc;
nc.id = CORBA::string_dup("Access");
nc.kind = CORBA::string_dup("Access");
SecurityDomain::Name nm;
nm.length(1);
nm[0] = nc;
```

Once the name component has been defined, it is put into the domain name list that was previously set up:

```
dl[0] = nm;
```

Finally, the mapping can be added to the ODM manager. The level of granularity is set to be "1" (i.e., the default (lowest) level of granularity).

```
dmanager1->
    set_domain_name_key((CORBA::UShort)1, okey2, dl);
```

Bank

Now we can set the mapping for the "Bank" domain. It maps the key <server host X.509 identity, /RootPOA/BankPOA/> to the domain "/Access/Bank" (i.e., the domain "Bank" used for access control). In this example, ODM maps at a per-POA granularity, so all objects within the scope of "/Root-POA/BankPOA" will be in the domain "/Access/Bank". Again, we first define the key for the mapping:

```
Security::Opaque okey3;
string key3 = "[/C=UK/ST=Server
    State/L=Cambridge/O=ObjectSecurity Ltd.
    /OU=RD/CN=Server
    Test/Email=server@test]/RootPOA/BankPOA";

int len3 = key3.length();
okey3.length(len3);
for (int i = 0; i < len3; i++)
    okey3[i] = key3[i];
SecurityDomain::NameList dl3;
dl3.length(1);
```

Then we define a single domain list with a domain name that consists of two name components, "Access" and "Bank":

```
SecurityDomain::NameComponent nc3;
nc3.id = CORBA::string_dup("Access");
nc3.kind = CORBA::string_dup("Access");
SecurityDomain::Name nm3;
nm3.length(2);
nm3[0] = nc3;
nc3.id = CORBA::string_dup("Bank");
nc3.kind = CORBA::string_dup("Access");
nm3[1] = nc3;
dl3[0] = nm3;
```

Then, we set the mapping of this key to the domain name at the medium granularity level "2" (i.e., per-POA granularity):

```
dmanager1->
    set_domain_name_key((CORBA::UShort)2, okey3, dl3);
```

Account

Setting the mapping for the Account objects is done in a similar fashion. All Account objects will be held within the scope of the "AccountPOA" and will be mapped onto the domain "Accounts" of type "Access."

```
Security::Opaque okey4;
string key4 = "[/C=UK/ST=Server
    State/L=Cambridge/O=ObjectSecurity Ltd.
    /OU=RD/CN=Server
    Test/Email=server@test]/RootPOA/AccountPOA";

int len4 = key4.length();
okey4.length(len4);
for (int i = 0; i < len4; i++)
    okey4[i] = key4[i];
SecurityDomain::NameList dl4;
dl4.length(1);

SecurityDomain::NameComponent nc4;
nc4.id = CORBA::string_dup("Access");
nc4.kind = CORBA::string_dup("Access");
SecurityDomain::Name nm4;
nm4.length(2);
nm4[0] = nc3;
nc4.id = CORBA::string_dup("Accounts");
nc4.kind = CORBA::string_dup("Access");
nm4[1] = nc3;
dl4[0] = nm3;
dmanager1->
    set_domain_name_key((CORBA::UShort)2, okey4, dl4);
```

Finally, the ODM mapping is saved into an ODM configuration file ODM.map. This way, the configuration is made persistent for future use.

```
factory->saveConfigFile("ODM.map");
```

6.5.4.5   Domain Manager's Hierarchy

In step three, we have to create a hierarchy of `DomainManager` objects that matches the POA hierarchy. For demonstration purposes, we separate this process from the previous one (i.e., we define all the domain names again as they are needed). It would also be possible to reuse the domain names defined above, but this way it becomes clearer which variables are used in what context.

To begin, we get an initial reference for the `DomainManagerFactory` and narrow it to `factobj`. It will be used to create the individual `DomainManager` objects:

```
CORBA::Object_var factobj =
    orb-> resolve_initial_references
        ("DomainManagerFactory");
SecurityDomain::DomainManagerFactory_var dmfactory =
    SecurityDomain::DomainManagerFactory::
    _narrow(factobj);
```

Default

We first define a root `DomainManager` `dmroot` of type "Access" for the default root domain and cast it to the correct type:

```
dmfactory-add_root_domain_manager("Access");
SecurityDomain::DomainManagerAdmin_ptr dmroot =
    dmfactory->get_root_domain_manager("Access");
SecurityDomain::DomainAuthorityAdmin_ptr daaroot =
    SecurityDomain::DomainAuthorityAdmin::
    _narrow(dmroot);
```

Bank

Now we have to create a `DomainManager` for the second level domain "Bank" of the type "Access":

```
SecurityDomain::DomainManagerAdmin_ptr ndo =
    dmfactory->create_domain_manager();
```

Then a domain name "Bank" (with type "Access") is defined and added to the `DomainManager`:

```
SecurityDomain::Name first;
first.length(1);
first[0].id = CORBA::string_dup("Bank");
first[0].kind = CORBA::string_dup("Access");


daaroot->add_domain_manager(ndo, first);
```

Account

The same has to be done for the "Accounts" domain (type "Access"):

```
SecurityDomain::DomainManagerAdmin_ptr ndo1 =
    dmfactory->create_domain_manager();
SecurityDomain::Name second;
second.length(1);
second[0].id = CORBA::string_dup("Accounts");
second[0].kind = CORBA::string_dup("Access");
daaroot->add_domain_manager(ndo1, second);
```

6.5.4.6 Required Rights

Step four involves setting the rights required to access the target objects. Although MICOSec supports the use of any single alphabetic characters as an access right, this example only uses the standardized four rights get g, set s, use u, and manage m.

Default

The default domain "Access" should automatically apply to any target object on the specified host that is not covered by any of the other, more specific domains (i.e., "Bank" or "Accounts"). In this example (as well as in most real-world application scenarios), objects in the default domain do not require any specific rights.

Bank

Target objects in the "Bank" access control domain should require specific access rights. "Access" is the root of the access control domain hierarchy, so we have to look for this domain. The domain name and the length have to be defined relative to that root. The name variable is then filled with the domain name specified in the variable first, which has been previously defined for the "Bank" DomainManager:

```
SecurityDomain::Name fullnameBank;
fullnameBank.length(1);
fullnameBank[0] = first[0];
```

Then we get the `DomainManager` for the defined full name of the "Bank" domain and narrow it to `dmath` for further use:

```
SecurityDomain::DomainManagerAdmin_ptr dm =
   daaroot->get_domain_manager(fullnameBank);
SecurityDomain::DomainAuthorityAdmin_ptr dmath =
   SecurityDomain::DomainAuthorityAdmin::_narrow(dm);
```

Setting the actual policy for the "Bank" domain involves a number of steps. First, we have to call the `DomainManager` associated with the "Bank" domain to get a reference to the `SecTargetInvocationAccess` policy that should be set:

```
CORBA::Policy_ptr po =
   dmath->  get_domain_policy
   (Security::SecTargetInvocationAccess);
SecurityAdmin::ObjectAccessPolicy_var polBank =
   SecurityAdmin::ObjectAccessPolicy::_narrow(po);
```

We would like to define that the standard manage right m is required to access the operation `create` on the `Bank` interface. Before that can be done, a rights list with a single list item has to be defined and filled with the manage m right:

```
Security::RightsList rlist1;
rlist1.length(1);
Security::Right right1;
right1.rights_family.family_definer = 0; // OMG
right1.rights_family.family = 1;         // corba

right1.rights_list = CORBA::string_dup("m");
rlist1[0] = right1;
```

Finally, the required right for this operation is set by invoking the operation `set_required_rights` on the `SecTargetInvocationAccess` policy object:

```
polBank->set_required_rights("create",
  "IDL:Bank:1.0", rlist1, Security::SecAnyRight);
```

Next, we do the same for the open operation on the Bank interface, but this time we specify that either use u or get g are required. SecAnyRight states that any one of the rights is sufficient (union), whereas SecAllRights states that all of the specified rights are required (intersection). We chose the union combinatory.

```
Security::RightsList rlist12;
rlist12.length(2);
rlist12[0] = right1;
rlist12[1] = right1;
rlist12[0].rights_list = CORBA::string_dup("u");
rlist12[1].rights_list = CORBA::string_dup("g");
polBank->set_required_rights("open",
  "IDL:Bank:1.0", rlist12, Security::SecAnyRight);
```

To make this new policy effective, it needs to be fed back into the Domain-Manager of the "Bank" domain:

```
dmath->set_domain_policy(polBank);
```

We also have to tell the DomainManager how this policy should be combined with other policies. The current version of MICOSec supports the Union policy combinator, which collects all policies from the domain hierarchy to the root, and the FirstFit combinator, which applies the first policy (of the type "access") found in the hierarchy. We chose the union combinatory.

```
dmath->set_policy_combinator
  (Security::SecTargetInvocationAccess,
  SecurityDomain::Union);
```

Accounts

The required access rights for the "Accounts" domain are set in the same way. Again, we first define its name relative to the root, using the second domain name already defined. Then we get a reference to the domain manager and retrieve a reference to the SecTargetInvocationAccess policy:

```
SecurityDomain::Name fullnameAccounts;
fullnameAccounts.length(1);
fullnameAccounts[0] = second[0];

SecurityDomain::DomainManagerAdmin_ptr dm2 =
   daaroot->get_domain_manager(fullnameAccounts);
SecurityDomain::DomainAuthorityAdmin_ptr dmath2 =
   SecurityDomain::DomainAuthorityAdmin::_narrow(dm2);
CORBA::Policy_ptr po2 = dmath->get_domain_policy
   (Security::SecTargetInvocationAccess);
SecurityAdmin::ObjectAccessPolicy_ptr polAccounts =
   SecurityAdmin::ObjectAccessPolicy::_narrow(po2);
```

Then we set the required rights for the deposit operation on the Account
interface. We specify a right that contains use and set, and set this right in
the access policy. The combinator is set such that either use or set will be
sufficient to access the operation:

```
Security::RightsList rlist2;
rlist2.length(2);
Security::Right right21;
right21.rights_family.family_definer = 0; // OMG
right21.rights_family.family = 1;          // corba
rlist2[0] = right21;
rlist2[1] = right21;
rlist2[0].rights_list = CORBA::string_dup("u");
rlist2[1].rights_list = CORBA::string_dup("s");

polAccounts->set_required_rights("deposit",
   "IDL:Account:1.0", rlist2, Security::SecAnyRight);
```

Access to the balance operation should require the use right:

```
Security::RightsList rlist3;
rlist3.length(1);
rlist3[0] = right1;
rlist3[0].rights_list = CORBA::string_dup("u");

polAccounts->set_required_rights("balance ",
   "IDL:Account:1.0", rlist3, Security::SecAnyRight);
```

To withdraw money, either use or get rights should be required:

```
rlist2[0] = right1;
rlist2[1] = right1;
rlist2[0].rights_list = CORBA::string_dup("u");
rlist2[1].rights_list = CORBA::string_dup("g");


polAccounts->set_required_rights("withdraw ",
   "IDL:Account:1.0", rlist2, Security::SecAnyRight);
```

Now the generated policy can be passed to MICOSec as follows:

```
dmath2->set_domain_policy(polAccounts);
dmath2->set_policy_combinator
   (Security::SecTargetInvocationAccess,
   SecurityDomain::Union);
```

6.5.4.7 Rights Granted to Principals

In this final step, rights have to be granted to the principals involved in this example application. The list of rights granted to all principals is kept in the SecurityManager object, so we first need to obtain a reference to the SecurityManager from the ORB:

```
CORBA::Object_var objsecman =
   orb-> resolve_initial_references
      ("SecurityManager");
SecurityLevel2::SecurityManager_var secman =
   SecurityLevel2::SecurityManager::
   _narrow(objsecman);
```

Then we get a reference to the access rights list associated with this SecurityManager:

```
SecurityLevel2::AccessRights_ptr ar =
   secman->access_rights();
```

This list of access rights specifies the rights that have been granted to each principal. In the access control model, principals are expressed by their AccessID attribute, and so the access rights have to be associated with the principal's AccessID. In MICOSec, the AccessID contains the X.509

distinguished name of the principal. It is also possible to express groups
of principals by using the X.509 OU identity. Such group identities can be
accessed through the `PrimaryGroupId` attribute.

Manager

In order to be able to grant access rights to the manager, we need to specify
the manager's `AccessID` security attribute. In MICOSec, this is done as follows. First, we define the manager's X.509 identity as string and copy it to
the attribute value of type `opaque`:

```
int i;
Security::SecAttribute attr1; // manager
attr1.attribute_type.attribute_family.family_definer
    = 0;
attr1.attribute_type.attribute_family.family = 1;
attr1.attribute_type.attribute_type =
    Security::AccessId;
string str = "/C=UK/ST=State/L=Cambridge/
    O=ObjectSecurity Ltd./OU=Section/CN=Manager
    /Email=manager@ObjectSecurity.com";
len = str.length();
attr1.value.length(len + 1);
for (i = 0; i < len; i++)
    attr1.value[i] = str[i];
attr1.value[len] = 0;
```

We also have to create a rights list with a single item, the `manage` right:

```
Security::RightsList rlist11;
Security::Right right2;
rlist11.length(1);                           // Manager
right2.rights_family.family_definer = 0; // OMG
right2.rights_family.family = 1;         // corba
right2.rights_list = CORBA::string_dup("m");
rlist11[0] = right1;
```

Finally, we can grant the `manage` right to the principal "manager":

```
ar->grant_rights
    (attr1, Security::SecInitiator, rlist11);
```

Owner and Wife

The principals "owner" and his "wife" should both belong to the same group "family," which contains the group's X.509 OU attribute. By using the X.509 OU attribute instead of the X.509 principal identity, it is possible to have separate principal identities for "owner" and "wife," but, at the same time, associate them with a common `PrimaryGroupId` "family."

```
Security::SecAttribute attr2;
```

First, we define the group "family":

```
attr2.attribute_type.attribute_family.family_definer
  = 0;
attr2.attribute_type.attribute_family.family = 1;
attr2.attribute_type.attribute_type =
  Security::PrimaryGroupId;
str = "family";
len = str.length();
attr2.value.length(len + 1);
for (i = 0; i < len; i++)
  attr2.value[i] = str[i];
attr2.value[len] = 0;

Security::RightsList rlist22;
```

Then we specify a standard right use:

```
rlist22.length(1);
right2.rights_family.family_definer = 0;  // OMG
right2.rights_family.family = 1;           // corba
right2.rights_list = CORBA::string_dup("u");
rlist22[0] = right2;
```

Finally, we can associate the specified use right with the defined group "family":

```
ar->grant_rights
  (attr2, Security::SecInitiator, rlist22);
```

After running through the five steps the ODM, the required rights of the domains, and the granted rights to the principals are set. Now both the ODM and access control will be automatically enforced by MICOSec whenever an invocation arrives.

The rest of the server code has not been modified. It creates a `Bank` object, activates it, and writes the reference into a file called `Bank.ref`. After that, the POA and ORB are activated and start serving requests. As usual for servers, the shutdown operation at the end is never reached.

```
Bank_impl * micocash =
    new Bank_impl(accountpoa);

PortableServer::ObjectId_var oid =
    bankpoa->activate_object (micocash);

ofstream of ("Bank.ref");
CORBA::Object_var ref =
    bankpoa->id_to_reference (oid.in());
CORBA::String_var strn =
    orb->object_to_string (ref.in());
of << strn.in() << endl;
of.close ();

printf ("Running.\n")

mgr->activate ();
orb->run();

bankpoa->destroy (TRUE, TRUE);
delete micocash;
return 0;
}
```

6.5.5 Client-Side Code Example

In this example (and most real-world applications), the access control is only enforced at the target side, and so the client application is not concerned with security. The only difference to the client applications previously described is that now the client has to be able to handle frequent exceptions passed back from the target whenever a request gets rejected by the access control system. The modified client application in this example simply catches such exceptions and continues. If the client were not modified to catch this exception, then it would crash every time a request is rejected by the target security system.

The example uses two client applications, one for the manager (`client.cc`) and one for the owner and his wife (`client2.cc`). We will briefly discuss each in turn.

6.5.5.1 Manager Client

The client application (`client.cc`) is used by the manager. It is supplied with the manager's X.509 certificate at the command line specified in the client shell script. The application code first initializes its underlying ORB:

```
#include "account.h"

#ifdef HAVE_UNISTD_H
#include <unistd.h>
#endif

int
main (int argc, char *argv[])
{
    cout << "Manager started" << endl;
    CORBA::ORB_var orb = CORBA::ORB_init (argc, argv);
```

Then it loads the IOR of the `Bank` target application, which was saved into the file `Bank.ref` (in the local directory) by the server:

```
char pwd[256], uri[300], uri2[300];
sprintf
    (uri, "file://%s/Bank.ref", getcwd(pwd, 256));
```

It converts this reference into an IOR object, narrows it, and tries to connect to the `Bank` target. The application exits if the connect is unsuccessful, otherwise it continues:

```
CORBA::Object_var obj = orb->string_to_object (uri);
Bank_var bank = Bank::_narrow (obj);

if (CORBA::is_nil (bank)) {
    printf ("oops: could not locate Bank\n");
    exit (1);
}
```

Now the manager can create an account by calling the `create` operation on the `Bank`. The `create` operation can only be accessed by managers and, thus, requires the manage m right. Again, the application exits if the account creation was unsuccessful; otherwise, it continues:

```
Account_var account;
try {
    account = bank->create ();
} catch (...) {
    cout << "Couldn't create an account!" << endl;
}

if (CORBA::is_nil (account)) {
    printf ("oops: account is nil\n");
    exit (1);
}
cout << "Manager: Account OK\n";
```

Now we can deposit and withdraw some money and query the balance. The application has to catch all exceptions raised by the invocation to make sure that the client does not crash if an invocation gets rejected by the server-side access control enforcement system.

```
try {
account->deposit (700);
} catch (...) {
    cout << "Couldn't deposit!" << endl;
}

try {
account->withdraw (450);
} catch (...) {
    cout << "Couldn't withdraw!" << endl;
}
CORBA::Long bal = -1;

try {
    bal = account->balance ();
} catch (...) {
    cout << "Couldn't get balance!" << endl;
}
```

```
    printf ("Balance is %ld.\n", bal);
    return 0;
}
```

6.5.5.2 Owner and Wife Client

The second client application (client2.cc) is run twice by the client shell script. The first time it is supplied with the owner's X.509 certificate, while the wife's certificate is used the second time. The code is almost identical to the previous client application; the only difference is that other operations are invoked this time.

```
#include "account.h"

#ifdef HAVE_UNISTD_H
#include <unistd.h>
#endif

int
main (int argc, char *argv[])
{
    cout << "Owner/Wife started" << endl;

    CORBA::ORB_var orb = CORBA::ORB_init (argc, argv);
```

This application also loads the IOR of the Bank target application:

```
char pwd[256], uri[300];
sprintf
    (uri, "file://%s/Bank.ref", getcwd(pwd, 256));
```

It connects to the Bank target as follows:

```
CORBA::Object_var obj = orb->string_to_object (uri);
Bank_var bank = Bank::_narrow (obj);

if (CORBA::is_nil (bank)) {
    printf ("oops: could not locate Bank\n");
    exit (1);
}
```

Then it tries to open an account, using the open operation on Bank instead of the create operation that was used in the previous client application. In this example, the only difference to the create operation is that it requires different access rights. The open operation can be accessed by principals who posses the use and get rights, while manage is required to invoke create.

```
Account_var account;
try {
  account = bank->open ();
} catch (...) {
  cout << "Couldn't open an account!" ,<< endl;
}

if (CORBA::is_nil (account)) {
  printf ("oops: account is nil\n");
  exit (1);
}
```

Finally, it also tries to deposit and withdraw some money and query the balance:

```
try {
account->deposit (700);
} catch (...) {
  cout << "Couldn't deposit!" << endl;
}
try {
account->withdraw (450);
} catch (...) {
  cout << "Couldn't withdraw!" << endl;
}

printf ("Balance is %ld.\n", account->balance ());

return 0;
}
```

6.6   Security Auditing

Security auditing makes users accountable for security-related actions and, thus, assists in the detection of actual or attempted security violations. This is achieved by recording details of security-relevant events in the system, such as principal authentication or object invocations. Security administrators can specify audit policies that describe which events should be audited and under which circumstances.

The CORBA security services specification conceptually separates security auditing from its nonrepudiation functionality, which is used at the application layer to create irrefutable evidence of the delivery or receipt of an invocation. Nonrepudiation evidence gets cryptographically linked to the principal identity when it is generated. Note that nonrepudiation is not part of level 2 security; instead, the specification describes it as an optional security facility. It will, therefore, not be covered at this point.

From a more technical perspective, auditing is not only useful for logging security relevant events but can also be used for debugging complex distributed systems as it produces a trail of all object invocations.

MICOSec's security auditing component supports selective auditing to three different audit storages: the standard Unix logging mechanism `syslog`, a flat file, and an SQL database (`PostgreSQL`). The generation of audit log entries is always done automatically at the ORB layer. The audit storage can be set either from the command line for security-unaware applications or through the level 2 audit interfaces for security-aware applications. Audit policies can only be set through the level 2 audit interfaces, from either the application or a security management tool. This section only covers the use of MICOSec's auditing functionality from the application layer—the use for security-unaware applications will be discussed in Chapter 7.

6.6.1   Audit Interfaces

The CORBA security auditing functionality mainly resides in the locally constrained objects `AuditDecision` and `AuditChannel`.

`AuditDecision` can be used to find out if an audit log should be created for a particular event. Internally, `AuditDecision` uses the information from the `AuditPolicy` to reach its decision. It also contains information about its associated audit storage, which is encapsulated in the `AuditChannel` object. The IDL interface for `AuditDecision` looks as follows:

```
interface AuditDecision { // Locality Constrained
   boolean audit_needed (
      in Security::AuditEventType event_type,
      in Security::SelectorValueList value_list
   );
   readonly attribute AuditChannel audit_channel;
};
```

IDL 2: Standard audit decision

MICOSec's extended AuditDecision interface includes an additional operation create, which can be used to set up the physical audit storage for the AuditChannel from within the application:

```
boolean create(
   in string arch_type,
   in string arch_name
   );
```

IDL 3: Audit decision extension

The first argument selects the type of audit storage, while the second specifies storage type depend details, such as the file name or database name. Table 6.3 summarizes how MICOSec auditing can be set up to use a flat file, UNIX syslog, or PostgreSQL database:

Table 6.3
Audit Storage Types

Storage Type	Storage Dependent Details	Description
file	AuditLog.log	Writes audit data to a flat file called AuditLog.log
syslog	<priority>	Writes audit data to the Unix syslog with the given priority
db	auditdb= objectsecurity	Writes the audit records to PostgreSQL and uses a database called objectsecurity. Additional arguments can be given as a comma-separated list

The AuditChannel object has an identifier and contains one operation audit_write, which generates a log entry in the corresponding audit storage. It accepts the information that has to be logged, such as:

- The event type;
- A list of credentials of the principal responsible for the event;
- The time the event occurred;
- A list of selector values;
- Event-specific data associated with the event type.

This is the IDL interface of the AuditChannel object:

```
interface AuditChannel {
  void audit_write (
      in Security::AuditEventType event_type,
      in CredentialsList creds,
      in Security::UtcT time,
      in Security::SelectorValueList descriptors,
      in any event_specific_data
  );
      readonly attribute Security::AuditChannelId
          audit_channel_id;
};
```

IDL 4: Audit channel

6.6.2 Audit Filtering Policies

MICOSec allows auditing of a wide range of security-related events, but logging all of them would soon produce a large and unmanageable amount of irrelevant data. It is therefore necessary to restrict the event generation to only the relevant subset of all auditable events.

To reduce the number of logged events, CORBA security uses audit filtering policies to specify the circumstances under which object invocation is audited, such as:

- Specified operations on objects;
- Failed operations;

- Use of certain operations during certain time intervals (e.g., during the weekend);
- Operations invoked by a certain principal;
- Any combinations of these.

These audit filters can be defined in the respective audit policies on the client side (SecClientInvocationAudit) and server side (SecTargetInvocationAudit). Both are based on the Security-Admin::AuditPolicy object, which provides operations to set, clear, replace, and retrieve audit selectors and to associate a particular audit channel with the policy:

```
interface AuditPolicy : CORBA::Policy {
  void set_audit_selectors (
      in CORBA::RepositoryId object_type,
      in Security::AuditEventTypeList events,
      in Security::SelectorValueList selectors,
      in Security::AuditCombinator audit_combinator
  );

  void clear_audit_selectors (
      in CORBA::RepositoryId object_type,
      in Security::AuditEventTypeList events
  );

  void replace_audit_selectors (
      in CORBA::RepositoryIdf object_type,
      in Security::AuditEventTypeList events,
      in Security::SelectorValueList selectors,
      in Security::AuditCombinator audit_combinator
  );

  void get_audit_selectors (
      in CORBA::RepositoryId object_type,
      in Security::AuditEventType event_type
      out Security::SelectorValueList selectors,
      out Security::AuditCombinator audit_combinator
  );
```

```
void set_audit_channel (
    in Security::AuditChannelId audit_channel_id
);
};
```

IDL 5: Audit Policy

Within the audit policy, administrators can specify which events should be logged and under which circumstances. This is done by combining so-called *audit selectors* with the required *audit event types*. The combination method is selected with a so-called AuditCombinator, which can be set either to require all (SecAllSelectors) or any (SecAnySelector) of the selectors to match. The specification defines the standard event types summarized in Table 6.4.

MICOSec currently only supports the audit event types that can be fully implemented within the security services implementation: Audit-PrincipalAuth, which logs principal authentication; AuditSession-Auth, which logs the TCP connect; AuditInvocation, which gets

Table 6.4
Standard Event Types

Event Type	Event Description
AuditAll	All supported event types
AuditPrincipalAuth	Principal authentication
AuditInvocation	Invocation
AuditSessionAuth	Session authentication
AuditAuthorization	Authorization
AuditSecEnvChange	Change of the security environment *(currently unsupported in MICOSec)*
AuditPolicyChange	Policy change *(currently unsupported in MICOSec)*
AuditObjectCreation	Object creation *(currently unsupported in MICOSec)*
AuditObjectDestruction	Object destruction *(currently unsupported in MICOSec)*
AuditNonRepudiation	Nonrepudiation *(currently unsupported in MICOSec)*

triggered when an object is invoked; and `AuditAuthorization`, which logs access control decisions. The other audit event types are not implemented, as they would require significant changes to the MICO ORB and object adapter, in particular, for the event types related to object creation and destruction.

For each event type, a number of audit attributes (so-called audit selectors) can be selected, such as the interface or operation name. Each selector has to be filled with the corresponding value, such as the actual operation name. Table 6.5 summarizes the different selector types and their values.

An audit log entry is only generated when the event type specified in the `AuditPolicy` occurs and when all or any (depending on the combinator) of the specified selector values match with the event. In the example in Section 6.6.3, the event type is "invocation" and the selector is "operation" with the value "`hello`" as the operation name.

6.6.3   Building and Running the Example

As with the other examples, the audit demo application, which can be found in the MICO subdirectory `/demo/security/audit-aware`, is compiled by running `make`. The server and client application are again launched through the small shell scripts `rss` and `rcs`. There is no significant output on the console, but after the invocation is completed, the file `server.log` contains a number of log entries for principal authentication, the binding, the invocation of the `hello` operation, and the disconnect.

Table 6.5
Standard Audit Policy

Selector Type	Value on `audit_write` or `audit_needed`	Selector Type Description
Interface name	`CORBA::RepositoryId`	Target interface repository ID
Object reference	`IOR`	Target object reference
Operation	`op_name`	Invoked operation
Initiator	`Credentials`	Credentials of the initiator of the invocation
Success/failure	`boolean`	Success/failure of the event
Time	`utc`	Time when the event occurred
Day of week	`DayOfTheWeek`	Day of the week on which audit is to be done

Each log entry contains the following information:

- Time and date;
- Event type;
- Interface name;
- Object reference;
- Operation name;
- Initiator credentials;
- Information regarding the success or failure of the logged event;
- Information regarding whether the log was generated by a server or a client.

As part of the MICOSec audit example, the following audit log entry is created, because we have defined the selectors to trigger the creation of audit log entries for invocations of the operation hello:

```
Fri Sep  9 20:57:15 2001
Event=[AuditInvocation]
InterfaceName=[IDL:Hello:1.0]
ObjectRef=[iioploc://ssl:inet:ul201.objectsecurity.com:
12466//131.111.218.202/1568/984171429/%5f0]
Operation=[hello]
Initiator=[/C=UK/ST=Client
State/L=Cambridge/O=ObjectSecurity Ltd./OU=RD/CN=Client
Test/Email=client@test]
SuccessFailure=[no_info]
clientserver=[server]
```

Server.log

6.6.4 Target Example

This subsection discusses the example program in the usual fashion to show how the AuditChannel, AuditDecision, and AuditPolicy objects are used in practice. Note that this example only focuses on the target side. The client application is not relevant here; it simply acts as a trigger for the target's auditing functionality.

The server source code can be logically divided into two parts. The first part shows how the AuditChannel object is initialized, whereas the second part describes how the selectors in the AuditPolicy can be set from within the application.

The code starts with an implementation of the servants that implement two trivial operations hello and olleh, which each print out their name on the standard output:

```
#include <iostream.h>
#include "hello.h"

class Hello_impl : virtual public POA_Hello {
public:
  void hello () {
     cout << "Start Servant hello\n";
     cout << "hello\n";
     cout << "End Servant hello\n";
  }
  void olleh () {
     cout << "Start Servant olleh\n";
     cout << "olleh\n";
     cout << "End Servant olleh\n";
  }
};
```

The first logical part of the server source code initializes the POA, the servants, and the SecurityManager object in the usual fashion:

```
int
main (int argc, char *argv[])
{

  CORBA::ORB_var orb = CORBA::ORB_init
     (argc, argv, "mico-local-orb");
  PortableServer::POA_var poa;
  CORBA::Object_var poaobj =
     orb-> resolve_initial_references ("RootPOA");
  poa = PortableServer::POA::_narrow (poaobj);
  PortableServer::POAManager_var mgr =
```

```
    poa-> the_POAManager();
Hello_impl * micohello = new Hello_impl;
PortableServer::ObjectId_var oid =
    poa-> activate_object (micohello);
CORBA::Object_var ref =
    poa-> id_to_reference (oid.in());
CORBA::String_var ref_str =
    orb-> object_to_string (ref.in());

CORBA::Object_var sm =
    orb-> resolve_initial_references
        ("SecurityManager");
SecurityLevel2::SecurityManager_var secman =
    SecurityLevel2::SecurityManager::_narrow(sm);
```

MICOSec's auditing is, like access control, based on security domains. Before the audit policies can be configured, it is necessary to define the mapping between the objects and their security domains. This is done exactly like in the access control example, just with a different name and type. For the sake of simplicity, we only define a single domain called "/Audit".

```
ObjectDomainMapping::ODM_var odm =
    ObjectDomainMapping::ODM::_narrow(objodm);
ObjectDomainMapping::Factory_var factory =
    odm->current(); // in case we already have some ODM

if (CORBA::is_nil(factory))
    factory = odm->create(); // we don't have ODM
poa->registerODMFactory(factory);

ObjectDomainMapping::Manager_ptr dmanager1 =
    poa->get_ODM();
Security::Opaque okey2;
string key2 = "[/C=UK/ST=Server State
            /L=Cambridge/O=ObjectSecurity Ltd.
            /OU=RD/CN=Server Test
            /Email=server@ObjectSecurity.com] /";
```

```
int len = key2.length();
okey2.length(len);
for (int i = 0; i < len; i++)
  okey2[i] = key2[i];


SecurityDomain::NameList dl;
dl.length(1);

SecurityDomain::NameComponent nc;
nc.id = CORBA::string_dup("Audit");
nc.kind = CORBA::string_dup("Audit");
SecurityDomain::Name nm;
nm.length(1);
nm[0] = nc;
dl[0] = nm;
dmanager1->set_domain_name_key
  ((CORBA::UShort)1, okey2, dl);

factory->saveConfigFile("ODM.map");

CORBA::Object_var factobj =
  orb->resolve_initial_references
      ("DomainManagerFactory");
SecurityDomain::DomainManagerFactory_var dmfactory =
  SecurityDomain::DomainManagerFactory::
  _narrow(factobj);

dmfactory->add_root_domain_manager("Audit");
SecurityDomain::DomainManagerAdmin_ptr dmroot =
  dmfactory->get_root_domain_manager("Audit");
SecurityDomain::DomainAuthorityAdmin_ptr daaroot =
  SecurityDomain::DomainAuthorityAdmin::
  _narrow(dmroot);
```

After that, it gets a pointer to the AuditDecision object from the SecurityManager and invokes the MICOSec-specific operation create on it to establish a file called server.log as a new physical audit storage:

```
SecurityLevel2::AuditDecision_ptr auditdes =
   secman-> audit_decision();
SecurityLevel2::AuditChannel_ptr old_channel =
   auditdes->audit_channel();

CORBA::Boolean audit =
   auditdes-> create("file", "server.log");
```

The next logical part of the server shows how to set an audit policy that makes sure that only invocations on the operation `hello` are logged. This involves the following steps: First, a reference to the `PolicyCurrent` has to be obtained. `PolicyCurrent` contains references to all target-side security policies:

```
CORBA::Object_var policy_current_obj =
   orb-> resolve_initial_references
      ("PolicyCurrent");
SecurityLevel2::PolicyCurrent_var policy_current =
   SecurityLevel2::PolicyCurrent::
   _narrow(policy_current_obj);
assert (!CORBA::is_nil (policy_current));
```

`PolicyCurrent` is then used to get a pointer to the target-side `Audit-Policy` (`SecTargetInvocationAudit`) by narrowing down the first item in the policy type sequence as follows:

```
CORBA::PolicyTypeSeq policy_types;
   policy_types.length(1);
   policy_types[0] = Security::SecTargetInvocationAudit;

CORBA::PolicyList * policies =
policy_current -> get_policy_overrides
   (policy_types);

CORBA::Policy_ptr policy = (*policies)[0];

SecurityAdmin::AuditPolicy_ptr audit_policy =
   SecurityAdmin::AuditPolicy::_narrow(policy);
```

Now the `AuditPolicy` object is available through the pointer `audit_` `policy`. It will be used to set the audit events, selectors, and combinators. But before that can be done, a list of the events to audit has to be created and filled with the right values. In our example, we would like to audit invocations, so we select the event type `AuditInvocation`:

```
Security::AuditEventTypeList events;
events.length(1);

Security::ExtensibleFamily family;
family.family_definer = 0;
family.family = 12;

events[0].event_family = family;
events[0].event_type = Security::AuditInvocation;
```

Then a list of selectors is created and filled with a list of the selector types that specify which events should be audited. For simplicity, this example only uses one selector type, the name of the invoked operation. As a selector value, we define the operation `hello` so that a log entry will be created whenever this operation is invoked:

```
int i = 0;
Security::SelectorValueList selectors;
selectors.length(0);

i++;
selectors.length(i+1);
selectors[i].selector = Security::Operation;
selectors[i].value <<= "hello";
i++;
selectors.length(i+1);
```

If more than one selector is specified, then the application has to decide whether the record should be written if all selectors are true (`SecAll-Selectors`) or if at least one of the selectors is true (`SecAnySelector`). In this example, we set the combinator such that all selectors must be true:

```
Security::AuditCombinator audit_combinator =
    Security::SecAllSelectors;
```

Then we can invoke the `set_audit_selectors` operation on `Audit-Policy` to set the audit policy for the specified event:

```
audit_policy->set_audit_selectors
    ("",events,selectors,audit_combinator );
```

Finally, we associate the audit policy with the audit domain:

```
daaroot->set_domain_policy(audit_policy);
```

The remainder of the server code activates the servants through the POA manager and runs the ORB:

```
cout << "Activate Servant\n";
mgr->activate ();
orb->run ();
return 0;
}
```

6.7 Delegation

6.7.1 Overview

The CORBA security services support delegation of `Credentials` for both security-unaware and security-aware applications. Delegation means that, if an initiator has invoked an intermediate, then this intermediate can invoke a target on behalf of the initiator. To indicate the initiator's identity to the target, the intermediate needs to pass some or all of the initiator's credentials (and possibly some of its own credentials) to the target. This can be repeated several times, so that the initiator's credentials are delegated to the final target through a chain of invocations.

The decision regarding which credentials are used when an intermediate object in a chain invokes another object is controlled by the delegation policy at each node. The `SecurityAdmin::DelegationPolicy` object has operations that allow security administrators to set and get the delegation mode, such as no delegation (i.e., use the intermediate's credentials), impersonation (i.e., use the initiator's credentials), and several compositions of the initiator's and intermediate's credentials.

For a security-unaware intermediate object, the specified delegation mode is automatically enforced by the CORBA security services (see

Figure 6.5). Whenever the intermediate object invokes a target on behalf of the initiator, the CORBA security services query the intermediate's delegation policy to find out whether the delegated credentials of the initiator or the own credentials of the intermediate (or a combination of both) should become the credentials to be used for the next invocation. If necessary, the intermediate object's principal will also be authenticated by the CORBA security services. The CORBA security services can retrieve the received initiator's credentials from the intermediate's Current object, combine it with the intermediate's own credentials, and set the resulting invocation credentials in Current. These credentials will then automatically be used for invocations from the intermediate to the target.

Security-aware applications can actively decide on a case-by-case basis which credentials are to be used when invoking further targets (see Figure 6.6). To do that, an application can use the get_attributes operation on Current to obtain the initiator's own credentials. After processing them according to the application's delegation requirements, it can put them back into Current (using set_attributes) to make them available for subsequent invocations. Applications are also able to specify whether credentials are to be used only at the target (e.g., for access control), or whether they can also be delegated further (including the delegation mode). It is important to

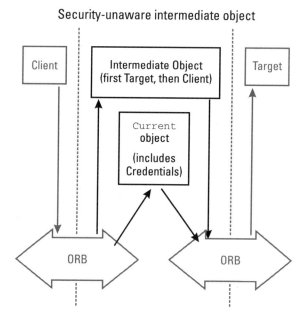

Figure 6.5 Security-unaware delegation.

Security-aware intermediate object

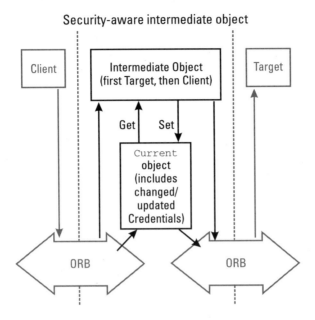

Figure 6.6 Security-aware delegation.

keep in mind that the features described here rely on the capability of the underlying security mechanisms to support delegation. In Section 6.7.2, we will discuss why CORBA security services implementations that are based on SSL cannot support delegation of credentials and how an additional security protocol layer can solve this problem.

6.7.2 Delegation Mechanisms

6.7.2.1 SSL Transport Layer Security

SSL is integrated into the CORBA security architecture as a security-enhanced underlying transport mechanism that provides (target-only or mutual) authentication and message encryption between two network sockets. Whenever a client ORB initiates an invocation, it opens an SSL connection to the target instead of a plain TCP/IP connection.

Each endpoint is associated with a cryptographic private key, and there is a publicly available X.509 identity certificate that links the corresponding public key to a principal name. This identity certificate is digitally signed by a certification authority that is trusted by both parties. The SSL authentication process verifies that the SSL implementation behind the remote socket has access to the private key that corresponds to the public key associated with the identity certificate. This mechanism works well as long as no delegation is required.

Now, if an intermediate would like to invoke another target on behalf of the initiator (i.e., impersonate the initiator), it would need to have access to the initiator's private key in order to authenticate successfully. However, this is not possible because the trustworthiness of the authentication process relies critically on the fact that private keys are only known to their respective owner and they are never revealed to anyone. As a result, SSL alone cannot support the delegation functionality described in the CORBA security services specification.

6.7.2.2   Security Attribute Service (CSIv2-SAS)

To overcome this (and other) weaknesses of the underlying secure transport, the OMG specified a security attribute service (SAS) as part of the *Common Secure Interoperability* (CSIv2) architecture [2]. The SAS resides on top of the underlying secure transport mechanism (typically SSL[6]) and provides client authentication, delegation, and privilege token functionality.

The SAS protocol is modeled after the GSSAPI token exchange paradigm [3] and exchanges its protocol elements in the GIOP service context. It consists of the following two layers:

- The higher *attribute layer* allows clients to transfer identity and privilege attribute tokens to a target where they may be applied in access control decisions. We will describe how this layer enables delegation over SSL.

- The lower *client authentication layer* can perform client authentication where sufficient authentication could not be accomplished by the underlying secure transport layer (e.g., when SSL is used). This functionality is not relevant to our discussion about delegation.

In essence, the SAS protocol allows tokens to be exchanged across a secure underlying transport. The X.509 identity tokens exchanged at the attribute layer allow an intermediate to act on behalf of (i.e., impersonate) some identity other than its own. To accept such a delegated identity, the target either has to trust the intermediate directly or base its trust on a proxy rule certificate (called authorization token [4]) that has been signed either by

---

6. Or SECIOP.

the initiator or a trusted privilege authority. Such a proxy certificate specifies whether or not the intermediate is authorized to act on the initiator's behalf.

Figure 6.7 illustrates the basic steps involved in the SAS delegation process, assuming SSL is used as an underlying secure transport mechanism.

1. When the initiator A invokes the intermediate node B, the underlying secure transport layer sets up a secure, authenticated SSL connection between A and B.

2. With the IIOP request from A to B, the SAS protocol then transfers an identity token (e.g., X.509) that links A's identity to its cryptographic public key.

3. Now the intermediate node B decides to invoke the target C on A's behalf, which establishes a secure SSL connection between B and C. At this point, C can only verify that B is on the other end.

4. To inform C that B's invocation has, in fact, been done on behalf of A, the SAS protocol then transfers A's identity token from B to C.

5. In addition, C needs a token (signed by A or a trusted privilege authority) that states that B is entitled to act on A's behalf. Using the public key from A's identity token (or the key associated with the privilege authority), C can verify the validity of the delegation and either accept or reject the request. This process is called *forward trust evaluation*, because the client (or the privilege authority) decides on the proxy rules. Alternatively, the target can evaluate its own set of proxy rules, which is called *backward trust evaluation*.

Although CSIv2 is security related, it is not specified within the CORBA security services specification but as part of CORBA v2.4. The current CORBA security services (v1.8. draft) do not provide any specific level 2

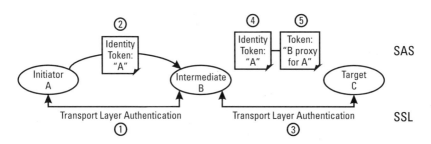

Figure 6.7 CSIv2-SAS delegation.

security interfaces to get or set CSIv2 tokens, but the existing `get_attributes` and `set_attributes` operations on `Current` can be used to insert and retrieve `Credentials` (which conceptually form the basis of SAS tokens).

At the time of this writing, the relatively new CSIv2 protocol is hardly used in practice, but it is likely that it will be widely used in the near future, first because it runs on top of the extremely widely-used SSL protocol, and secondly because it is supported by both CORBA and EJB and, thus, enables secure interoperability between both technologies. Currently, no implementation of CSIv2 is available for the standard MICO or MICO-Sec distribution, but it is anticipated that one will become available in the near future.

6.8   Implementation Overview and Conformance

Level 2 principal authentication and security context establishment are implemented in the same way as previously described for level 1 security (see Section 5.5). In fact, both level 1 and level 2 implementations are part of a single MICOSec implementation, which reuses the security functionality automatically provided by SSL whenever a transport connection is established.

One of the main design differences between both conformance levels is that level 2 uses the more flexible `Credentials` model to represent security attributes to the application layer. In line with level 2 conformance requirements, it is also possible to choose the quality of protection from the application layer (by selecting the SSL cipher suite), to change the privileges in `Credentials` object and to choose which credentials are to be used for object invocations (by setting them as the invocation credentials). SSL also supports the required peer authentication and message protection at (or rather below) the ORB layer and protection from replay/reorder attacks. Most other SSL-related implementation details are cumbersome and irrelevant to the user, and will therefore not be discussed here. For example, SSL policies such as `TrustInClient` have to be set up within the SSL implementation to conform to level 2 security.

Principal authentication from within the application is done the same way as for level 1. The application either calls the `PrincipalAuthenticator` with an identity certificate as a parameter, or a certificate is supplied to MICOSec as a command line argument.

Although the ODM is of critical importance to MICOSec's access control and audit functionality, it is currently not part of security level 2. As a consequence, ODM and its modifications to MICO's ORB and POA implementation, as well as changes in the CORBA access control and audit models, are not covered by level 2 conformance requirements. However, MICOSec's ODM is based on the SDMM [1], which should become an adopted OMG standard in the near future.

One of the goals of the ODM architecture was to keep modifications to the CORBA specification as minimal as possible to fulfill CORBA's portability and replaceability requirements. A few modifications to the MICO ORB were necessary, in particular two functions to the POA (`register-ODMFactory` and `get_ODM`), which allow the registration of an ODM factory with a POA and to query the ODM. Also, an initial reference "`ODM`" had to be added, so that a reference to the ODM can be obtained from within interceptors and servants. Other internal details of ODM are MICO-specific and subject to frequent change.

Although conceptually separate, access control and audit (as well as the ODM) are implemented inside a single MICO-specific interceptor so-called the `AuditInterceptor`. This is mainly done for MICO-specific optimization reasons. Whenever a message arrives at the ORB layer, the relevant access control and audit policies are evaluated and enforced automatically inside this interceptor. As part of this process, the interceptor implementation obtains the necessary data (e.g., security attributes) from the SSL transport layer and the request header from the ORB and compares it to the policy. Based on the policy, it then reaches a decision and enforces it accordingly. At this level of abstraction, everything is implemented close to the specification. Internally, there are more implementation details that are not directly relevant to the user[7] and which are subject to frequent change in order to keep up with changes in the MICO ORB. Both access control and audit policies use ODM domain names to describe the target objects the policy should apply to in an unchanging and human-readable way.

For `Credentials` delegation, MICOSec relies on the implementation of the SAS, which allows security tokens to be sent over a secure transport connection. These token layers enable initiators (or trusted third parties) to grant intermediates the right to act on behalf of the initiator. The SAS can be integrated into MICOSec without any changes to the application-facing level 2 interfaces, because all tokens can be easily accessed through the existing `Current/SecurityManager` interface.

---

7. For example, the Standard Template Library (STL) is used to search the rights lists.

6.9 Summary

This chapter covers the use of the application-facing CORBA security level 2 interfaces. For demonstration purposes, the bank application from Chapter 1 is modified several times to make use of different security level 2 interfaces.

Level 2 security incorporates a rich set of application-facing interfaces to make use of the full set of the security facilities described in the CORBA security services specification. The level 2 interfaces allow applications to administer fine-grained policies to control the security provided at object invocation. While the level 2 functionality and interfaces are richer and more flexible than those at level 1, they are, at the same time, more complex in their use, in particular the interfaces related to security associations.

The level 2 security associations are based on the more flexible concept of Credentials objects instead of Current. Credentials are normally established during principal authentication and hold the security attributes of local and remote principals (such as the X.509 certificate identity) and fine-grained security association policies. The Bank authentication example illustrates how principal authentication and security association establishment are used in practice. It shows the use of the SecurityManager, Credentials, and PrincipalAuthenticator interfaces and how security attributes can be obtained from the application layer.

To allow administrators to express target objects in a scalable, unchanging, well-defined, and human-readable way, MICOSec's ODM maps "transient" target addressing information (such as the object reference or security attributes) onto persistent and "unchanging" domain names. This allows administrators to easily associate access control and audit policies with the target objects (or operations) to which the policy should apply.

MICOSec's access control functionality uses ODM domain names to associate access policies with the targets. The Bank access control example demonstrates the use of the objects AccessDecision, Required-Rights, and DomainManager, which are used to bootstrap the ORB layer access control enforcement and configure the access policy. All access control-related modifications to the example code are in the server, so that the servant implementation remains unmodified (i.e., security-unaware). The example involves three principals with different access rights for different target operations and demonstrates how invocations are granted or rejected (based on the access policy).

MICOSec's security auditing component supports selective audit log generation into three different audit storages: the standard Unix logging mechanism syslog, a flat file, and an SQL database (PostgreSQL). The generation of audit log entries is always done automatically at the ORB layer. The simple auditing example shows how the audit storage and audit policies can be set through level 2 audit interfaces `AuditDecision`, `Audit-Channel`, and `AuditPolicy`. Again, audit policies are associated with the invoked target by using ODM domain names.

Delegation of `Credentials` cannot be supported by CORBA security services implementations that only use SSL transport layer security contexts to set up ORB layer security associations. On the SSL layer, an endpoint is identified by a public key certificate that corresponds to the private key used to encrypt the channel between two endpoints. The trustworthiness of the whole process relies critically on the fact that private keys are never revealed, so that the use of the private key proves the identity associated with the key. Now, if an intermediate target node wants to take on the initiator's identity (i.e., impersonate the initiator) when it invokes another target, then it would need to have access to the initiator's private key. To enable delegation, the OMG specified a SAS as part of the CSIv2 architecture [2]. The SAS resides on top of the underlying secure transport mechanism and provides an extra protocol layer that can be used to transfer tokens. This allows initiators to grant to other identities the right to be an intermediate and enables targets to verify that an intermediate node has been endorsed by the initiator.

6.10 Further Reading

There is no other literature apart from this book that is related to the use of CORBA security level 2 interfaces. Specific details on interfaces and conformance can be found in the CORBA security services specification [5]. Unfortunately, the specification's current version is not very readable and does not give any explicit use guidelines. The *MICOSec User's Guide* [6] contains a limited amount of documentation on MICOSec's level 2 interfaces and its use but does not go beyond what is already described in this chapter.

References

[1]   OMG, *CORBA Security Domain Membership Management Service*, Final Submission, July 2001.

[2]   OMG, *Common Secure Interoperability V2 Specification*, 2000.

[3]   Linn, J., *Generic Security Service Application Program Interface Version 2, Update 1*,(IETF RFC 2743), March 2001.

[4]   Farrell, S., and R. Housley, *An Internet Attribute Certificate Profile for Authorization*, (IETF ID PKIXAC), 2000.

[5]   OMG, *CORBA Security Services Specification*, Version 1.8 (Draft Adopted Revision), 2000.

[6]   Schreiner, R., and U. Lang, *MICOSec User's Guide*, ObjectSecurity Ltd., 2000, http://www.microsec.org.

7

Security-Unaware Functionality

7.1 Introduction

In Chapters 5 and 6, you learned how to use the CORBA security services from within your security-aware application. Applications can use the level 1 or level 2 application-facing interfaces for two purposes: to configure the policy (and other options) for the underlying ORB layer security functionality and to obtain security-related information from the ORB layer for fine-grained security enforcement within the application.

But the CORBA security services architecture also caters to security-unaware applications. These applications should not contain any security-related code; instead, all security configuration is done by other means. In MICOSec, various security configuration parameters have to be supplied as command line arguments when the application is launched. Some of the arguments also specify configuration files (e.g., for object domain mapping, access control, and audit).

Although less flexible than application layer security, ORB layer security for security-unaware applications has a number of advantages. First, it allows the separation of the application development task from the security system implementation. This way, application developers do not need to know anything about security and security policy, which allows them to focus on the actual application development process. Secondly, it allows security administrators to configure the automatic enforcement of security policies that can be managed without any active involvement of the application developer or user.

This chapter describes the configuration of MICOSec's functionality for security-unaware applications. This involves command line arguments to set up SSL parameters and X.509 identity certificates for principal authentication and security association establishment, as well as configuration files for object domain mapping, access control policies, principal rights, and audit filters.

Section 7.2 summarizes the main functional parts that have to be supported at the ORB layer for level 1 and level 2 conformance. The subsequent sections then describe the command line arguments and configuration files required to set up MICOSec for security-unaware applications.

7.2 Security-Unaware Functionality Overview

The CORBA security services specification [1] defines which security features have to be provided at the ORB layer to security-unaware applications (i.e., without the need for any security-related modifications to the application source code). The specified functionality includes support for the configuration of security features and security policies, as well as the actual evaluation and enforcement of security policies.

The range of supported ORB layer security features depends on the conformance level. Level 1 conformant products need to provide the following functionality at the ORB layer:

- Principal authentication inside *or* outside the object system;
- Secure invocation between client and target object (including unilateral authentication, integrity, and/or confidentiality) on the ORB layer *or* outside the object system;
- Simple delegation of client security attributes to targets, depending on the supported CSI level;
- ORB-enforced access control checks, with support for domains and roles but no support for administration;
- Auditing of security-relevant system events (but not by object invocation).

For level 2 conformance, security services need to support extra ORB layer functionality on top of the level 1 functionality:

- Principal authentication both inside and outside the object system;

- Additional secure invocation features, in particular, peer authentication and message protection at the ORB level;

- Further integrity options, such as replay/reorder protection (can be requested, but need not be supported by all implementations);

- Access control and selective auditing have to support a per-operation granularity.

As far as the functionality for security-unaware applications is concerned, level 1 defines a strict subset of level 2. We discussed in Chapters 5 and 6 that the internal architecture differs between the two conformance levels, in particular with respect to the `Credentials` model. But since there are no application-facing interfaces for security-unaware applications, these differences are not visible outside the ORB layer.

The following sections illustrate the configuration of MICOSec's level 2 conformant features for security-unaware applications.

7.3 Principal Authentication and Secure Association

The principal authentication process associates an application object with a principal identity (e.g., the user's identity) and makes the resulting credentials available for use during security association establishment. In MICO-Sec, a principal identity consists of an X.509 identity certificate and a corresponding key pair. To configure the SSL transport layer, both the certificate and key file have to be supplied to MICOSec as an additional command line argument when the application is launched.

Once the SSL parameters have been set up, the SSL transport layer automatically establishes the security association whenever an invocation occurs, unless a security association to the target has already been previously set up by an earlier invocation.

7.3.1 Command Line Arguments

MICOSec uses the following MICO SSL command line arguments for principal authentication (i.e., to specify the files that contain the X.509 certificate and cryptographic key):

- `ORBSSLcert <certificate file>`
  This command line option specifies the file that holds the X.509

certificate for the launched client or target application. `OpenSSL` files use the extension `.pem` for key and certificate files. This argument defaults to `default.pem`.

- `ORBSSLkey <key file>`
  This option specifies the `.pem` file that holds the key pair for the launched client or target application. It defaults to the same file as the certificate file.

The SSL automatically establishes the security association whenever a new transport connection is established. As part of this process, the certificates are transferred securely to the remote peer, and optionally checked for validity with a certification authority. Also, the ciphers that should be used during the association are negotiated. The following MICO SSL command line arguments configure MICOSec's ORB layer security association features:

- `ORBSSLcipher <colon separated list of preferred ciphers>`
  This parameter can be used to specify the ciphers that the launched client or target is willing to support. If it is not specified, then an implementation-specific default policy is used instead, which depends on the cryptographic functions supported by the specific implementation, as well as on cryptography export regulations and patents in some countries.
  Commonly used cipher suites include: NULL-MD5, RC4-MD5, EXP-RC4-MD5, IDEA-CBC-MD5, RC2-CBC-MD5, EXP-RC2-CBC-MD5, DES-CBC-MD5, DES-CBC-SHA, DES-CBC3-MD5, DES-CBC3-SHA, and DES-CFB-M1.

- `ORBSSLverify <verify depth>`
  If this parameter is specified, then the peer must supply a valid certificate, otherwise the connection setup will fail. `<verify depth>` specifies how many hops of the chain of certification authorities should be checked. By default, the validity of the peer certificate is not checked.

- `ORBSSLCAfile <CA filename>`
  This argument specifies the `.pem` file that holds the certificates of certificate authorities (CA).

- ORBSSLCApath <CA pathname>

  This parameter can point to the directory that contains .pem files holding certificates of CAs. It defaults to /usr/local/ssl/certs.

7.3.2 Example Configuration

The security-unaware access control example (which can be found in the MICOSec subdirectory /demo/security/acl-unaware) demonstrates the configuration of the SSL parameters related to principal authentication and security association establishment. The application functionality is identical to the security-aware access control example described in Section 6.5, but this time the application code does not contain any security-related code.

The server and client start-up shell scripts show the use of the configuration parameters described in this section:

```
./server
-ORBIIOPAddr ssl:inet:'uname -n':12466
-ORBSSLcert ServerCert.pem
-ORBSSLkey ServerKey.pem
-ORBSSLverify 0
...
```

Server shell script

As described in Section 6.5, the client shell script starts three clients for the manager, the owner, and the owner's wife. A different X.509 identity certificate is supplied each time. For the sake of simplicity, the same key pair is used for all invocations. In most real-world environments, each principal would have a separate key pair, so that no user can eavesdrop on communications by other users.

```
#!/bin/sh

ADDR=ssl:inet:'uname -n':12456

echo "Manager"
./client
-ORBBindAddr $ADDR
-ORBSSLcert manager.pem
-ORBSSLkey key.pem
```

```
-ORBSSLverify 0

echo "Owner"
./client2
-ORBBindAddr $ADDR
-ORBSSLcert owner.pem
-ORBSSLkey key.pem
-ORBSSLverify 0

echo "Wife"
./client2
-ORBBindAddr $ADDR
-ORBSSLcert wife.pem
-ORBSSLkey key.pem
-ORBSSLverify 0
```

Client shell script

7.4   Object Domain Mapping

MICOSec's object domain mapping feature provides unchanging domain names that can be used to describe target objects inside access control and audit policies. Its functionality is based on the object domain mapper described as part of the SDMM [2]. In MICOSec, the ODM maps X.509 identities and POA names (and optionally object identifiers) onto hierarchical domain names.

Once the application program has been launched, the supplied ODM configuration is set up inside MICOSec. Whenever an invocation arrives at the target side, the mapping is automatically carried out to locate the access control and audit policies associated with the invoked target object.

The configuration file is supplied to MICOSec with the following command line argument:

- ODMConfig <config file>
  This argument specifies the .cnf file that holds the ODM mapping table.

The content of the configuration file defines mappings from X.509 identities and POA names (and optionally object identifiers) to hierarchical domain names. The following example ODM configuration file is part of the security-unaware access control example, which can be found in the MICO-Sec subfolder `/demo/security/acl-unaware`. It configures the domain mappings both for access control ("Bank" and "Account") and audit ("d1"):

```
[ /C=UK/ST=Server State/L=Cambridge/O=ObjectSecurity
    Ltd./OU=RD/CN=Server Test/Email=server@test ]
    /
    /Access

[ /C=UK/ST=Server State/L=Cambridge/O=ObjectSecurity
    Ltd./OU=RD/CN=Server Test/Email=server@test ]
    /
    /Audit

[ /C=UK/ST=Server State/L=Cambridge/O=ObjectSecurity
    Ltd./OU=RD/CN=Server Test/Email=server@test ]
    /RootPOA/MyPOA/
    /Access/Bank

[ /C=UK/ST=Server State/L=Cambridge/O=ObjectSecurity
    Ltd./OU=RD/CN=Server Test/Email=server@test ]
    /RootPOA/AccountPOA/
    /Access/Accounts

[ /C=UK/ST=Server State/L=Cambridge/O=ObjectSecurity
    Ltd./OU=RD/CN=Server Test/Email=server@test ]
    /RootPOA/AccountPOA/
    /Audit/d1
```

Configuration file for object domain mapping (ODM.cnf)

The application code remains unaffected by the ODM configuration and mapping processes. However, it is still necessary to manually establish a POA hierarchy that reflects the domain name hierarchy defined in the ODM configuration file. This is done by calling the `create_POA` operation.

```
PortableServer::POA_var mypoa =
    poa->create_POA ("MyPOA", mgr, pl);
mypoa->registerODMFactory(factory);
```

276 Developing Secure Distributed Systems with CORBA

In many complex real-world applications, there is already an existing POA hierarchy that reflects the business logic, because different objects often require different POA policies. This existing POA hierarchy can simply be reused for ODM. For security-unaware applications, MICOSec automatically registers the POA in the ODM.

7.5 Access Control

Access control is about restricting access to target objects. Access control enforcement can either be done at the application layer for security-aware applications or at the ORB layer for security-unaware applications. The configuration of MICOSec's ORB layer access control features can be done in two ways, either from within the application (by using the security level 2 interfaces) or with a number of configuration files, which need to be specified as command line arguments when the application is launched.

7.5.1 Bank Example

The security-unaware access control example can be found in the MICOSec subfolder /demo/security/acl-unaware. It has the same functionality as the security-aware access control example described in Section 6.5, but this time the rights are not set from within the application. Instead, two configuration files specify the rights granted to principals and the rights required to access target objects.

The example consists of the usual bank account application, which contains the Account interface (with the usual operations deposit, withdraw, and balance) and the Bank interface (with two operations create and open). It involves three client principals and two target objects with five operations. Each principal has different access rights for each target operation.

On the target host, the ODM groups Bank objects into the domain "Bank" and Account objects into the domain "Accounts." The target host is expressed by a certificate, and the domains are represented by the POA hierarchy:

```
[ /C=UK/ST=Server State/L=Cambridge/O=ObjectSecurity
     Ltd./OU=RD/CN=Server Test
     /Email=server@ObjectSecurity.com ]
  /
```

```
/Access

...

[ /C=UK/ST=Server State/L=Cambridge/O=ObjectSecurity
    Ltd./OU=RD/CN=Server Test
    /Email= server@ObjectSecurity.com ]
/RootPOA/MyPOA/
/Access/Bank

[ /C=UK/ST=Server State/L=Cambridge/O=ObjectSecurity
    Ltd./OU=RD/CN=Server Test
    /Email= server@ObjectSecurity.com ]
/RootPOA/AccountPOA/
/Access/Accounts

...
```

Configuration file for object domain mapping (ODM.cnf)

In addition to the ODM configuration file, ORB layer access control requires two other configuration files: the granted rights file and the required rights file.

The granted rights file defines the rights granted to the different principals in the system. The rights granted to the manager and owner/wife principal identities are summarized in Table 7.1.

In the configuration file, this information is represented as follows: First, the manager's principal identity (AccessId) is represented as the X.509 identity, followed by the manage right corba:m. Next, the access right use (corba:u) is associated with the group (GroupId) "family," which is stored as the OU attribute in the X.509 certificate of the owner and the wife:

Table 7.1
Granted Rights

Security Attribute	Attribute Value (Identities)	Granted Right
AccessID	Manager	Manage
PrimaryGroupID "family"	Owner and wife	Use

```
# Manager

[/C=UK/ST=State/L=Cambridge/O=ObjectSecurity
Ltd./OU=Section/CN=Manager/Email=manager@Object
Security.com]AccessId:corba:m

# Owner and wife

[family]GroupId:corba:u
```

Granted rights configuration file (rights.cnf)

The configuration file name can be supplied to MICOSec through the following command line argument:

- `RightsConfig <config file>`

  This argument specifies the `.cnf` file that holds the granted rights table.

Table 7.2 summarizes the content of the required rights configuration file. It shows which interface and operations require which rights, and into which domain the objects of this type are placed.

This information is expressed in the corresponding configuration file as follows: The required rights for the domain "/Access/Bank" and "/Access/Accounts" are each defined in a substructure that specifies the interface, the operation name, and the required right. In addition, the rights and policy combinators are specified for each case:

Table 7.2
Required Rights

Type/Domain (Policy Combinator union)	Interface	Operation	Required Rights (Rights Combinator any)
/Access/Bank	Bank	create	Manage
/Access/Bank	Bank	open	Use, get
/Access/Accounts	Account	deposit	Use, set
/Access/Accounts	Account	withdraw	Use, get
/Access/Accounts	Account	balance	Use

```
# Bank operations

/Access/Bank Combinator = Union
{
   IDL:Bank:1.0 create   corba:SecAnyRight:m
   IDL:Bank:1.0 open     corba:SecAnyRight:ug
}

#Account operations

/Access/Accounts Combinator = Union
{
   IDL:Account:1.0 deposit  corba:SecAnyRight:us
   IDL:Account:1.0 withdraw corba:SecAnyRight:ug
   IDL:Account:1.0 balance  corba:SecAnyRight:u
}
```

Required rights configuration file (access.cnf)

In addition to the four standard access rights (g, s, u, m), MICOSec supports all lowercase and uppercase characters as rights. It also supports two *meta rights*, which are useful to express general security policies:

- "*": No rights are required to invoke this operation; everybody is allowed to invoke it.

- "-": No effective right matches this meta right; nobody is allowed to invoke the operation.

The policy combinators specify how policies are collected in the domain hierarchy:

- "Union": The access control mechanism iterates through the whole domain tree, collects the required rights, and combines them.

- "FirstFit": The access control searches in the domain tree for the first occurrence of a matching access policy, and applies it. All other policies on the way toward the root are ignored.

The configuration file name can be supplied to MICOSec with the following command line argument:

- `AccessConfig <config file>`
  This argument specifies the `.cnf` file that holds the required rights table.

In addition, it is possible to set the following flags at the command line:

- `Paranoid <yes, no>`
  A default policy has to be applied if there is no entry for an operation in the list of required rights. This flag defines the default access control policy. If the flag is set to "`yes`", the default policy is to deny access, analogous to the "`-`" meta right. If it is set to "`no`", invocations of this operation are allowed, analogous to the "`*`" meta right.

- `AccessControl <on, off>`
  This flag is used to enable or disable the access control as a whole. It can be used if access control is not needed, but the administrator does not want to redefine the configuration files to permit everything.

The security-unaware access control example comes with the usual shell scripts: `rss` for the server and `rcs` for the client. Both command line arguments related to access control are included in the server shell script and specify the configuration files described above (in this example `access.cnf` and `rights.cnf`) to be used.

7.6 Security Auditing

Security auditing makes users accountable for security-related actions and, thus, assists in the detection of actual or attempted security violations. This is achieved by recording details (but not irrefutable evidence) of security-relevant events in the system, such as principal authentication or object invocations. Security administrators can specify audit policies that describe which events should be audited and under which circumstances.

MICOSec's security auditing component supports selective auditing to three different audit storages: the standard Unix logging mechanism `syslog`, a flat file, and an SQL database (`PostgreSQL`). The generation of audit log entries is always done automatically at the ORB layer. The audit storage can be set either from the command line for security-unaware applications or through the level 2 audit interfaces for security-aware applications. This section covers the configuration of MICOSec's auditing functionality for security-unaware applications.

Security-unaware auditing is enabled by setting the audit channels at the command line, without modifying the application source code. The options for the different channels, file, syslog, or database are:

- `AuditType` selects the audit channel; the arguments are:
  - `file`   Write audit data to a flat file.
  - `syslog` Write audit data to the Unix `syslog`.
  - `db`   Write audit data to the `PostgreSQL` database.

- `AuditArchName` specifies audit channel dependent details for the chosen audit name, such as the file name, syslog priority, or database name.

The following examples show the use of these command line options:

- `-AuditType db -AuditArchName dbname=objectsecurity` writes the audit records to `PostgreSQL` and uses a database "objectsecurity". Additional arguments can be given as a comma-separated list.
- `-AuditType syslog -AuditArchName 1` writes the records to the Unix `syslog` with the given priority 1.
- `-AuditType file -AuditArchName AuditLog.log` writes the records to the file `AuditLog.log`.

7.6.1 Example Configuration

The security-unaware audit example in the MICOSec subdirectory `demo/security/audit-unaware` illustrates how the audit filters are configured from the command line and with an audit filter configuration file. The client applications `Client` and `Client2` are identical to the security-aware access control example described in Section 6.5. The server code is identical to the code described in Section 7.5 for the access control example.

The client shell script `rcs` only sets up the parameters related to SSL keys and certificates, as no auditing will be done on the client side. The server shell script `rss` does the same, but then also specifies the filename of the

ODM configuration file to be used. In addition, it contains the following audit-related command line arguments:

```
./server
…
-AuditConfig audit.cnf
…
-AuditType file
-AuditArchName server.log
```

Server shell script

The file specifies the file that contains the audit configuration (audit.cnf), the type of audit channel (file), and the audit channel file name (server.log).

The audit configuration file contains the event filter specifications for two audit domains, the root "/Audit" and one branch "/Audit/d1." The root domain contains the most general default audit filter policy. It will apply to all events to which no more specific policies apply. For the domain "/Audit/d1", the configuration file audit.cnf specifies a more specific audit policy:

```
/Audit
{
(
 server:Any * [PrincipalAuth, SessionAuth]
)
}

/Audit/d1
{

(
 server:Any IDL:Account:1.0 [Authorization]
 Initiator = [ /C=UK/ST=Cambridge/L=Cambridge/
     O=ObjectSecurity/OU=Domain1/CN=manager/
     Email=manager@objectsecurity.com ]
 DayOfWeek = [Wed ]
 Time = [2001/05/29:11:15:00-2001/05/30:16:00:00]
```

```
SuccessFailure = [ false]
)


(server:All IDL:Bank:1.0 [Invocation]
Initiator = [ /C=UK/ST=Cambridge/L=Cambridge/
     O=ObjectSecurity/OU=Domain1/CN=manager/
     Email=manager@objectsecurity.com ]
  DayOfWeek = [Wed]
)


}
```

Audit configuration file (audit.cnf)

To express the audit policy, we first state the name of the domain branch. In our example, the root name is "/Audit", whereas the branch domain is "/Audit/d1". Next, audit filters are defined inside the " { } " brackets, and each individual audit filter specification is surrounded by " ( ) ".

The prefix "server" means that the audit event has to occur on the server side audit (MICOSec currently does not support client-side audit). "Any" stands for the audit combinator SecAnySelector, whereas "All" means SecAllSelectors. These values determine how many of the specified audit selectors need to match the event in order to trigger the creation of an audit log entry.

Next, the interface name follows. In the case of the default audit policy, " * " means that the filter should apply to all interfaces.

Inside the brackets " [ ] ", we can specify then the list of audit event types we are interested in. Possible values are shown in Table 7.3.

Table 7.3
Audit Events

Audit Event Types
All
PrincipalAuth
SessionAuth
Authorization
Invocation

The audit event types "Invocation" and "Authorization" are related to specific interfaces, whereas the other event types, "PrincipalAuth" and "SessionAuth," are interface independent.

The root audit policy specified in this example will match all events of the two specified types ("PrincipalAuth" and "SessionAuth"), because no specific audit selectors have been supplied.

The branch "/Audit/d1" defines a much more fine-grained audit policy with two filters for the different interfaces "IDL:Account:1.0" and "IDL:Bank:1.0." Filters can be defined using one of the selectors in Table 7.4.

Again, it would be possible to describe a default policy for this audit domain, using "*" as an interface name. This way, it is possible to cover all audit events for which no audit filters are explicitly specified. If no event and event parameters match any of the filters defined for the specified domain, then this default policy will be tested. If there is also no match, then the parent domain will be tested in turn, potentially until the default policy of the root domain is reached.

To be able to make use of audit domains, we also need to map an appropriate POA hierarchy onto appropriate branches of the audit domain hierarchy in the ODM configuration file. In this example, the file config.cnf defines mappings for the two audit domains as follows:

Table 7.4
Audit Selectors

Audit Selector	Value
Initiator	Credentials attribute AuditId
DayOfWeek	Mon, Tue, Wed, Thu, Fri, Sat, Sun
SuccessFailure	false, true
Time	yyyy/mm/dd:hh:mm:ss-yyyy/mm/dd:hh:mm:ss, (i.e., time interval—without any blank spaces)
Operation	Any interface operation or pseudo operation (e.g., _connect, _disconnect, _principalauth)

```
# Configuration file for Domain mapping

...


[/C=UK/ST=Server State/L=Cambridge/O=ObjectSecurity
Ltd./OU=RD/CN=Server Test/Email=server@test]/
/Audit


...

[/C=UK/ST=Server State/L=Cambridge/O=ObjectSecurity
Ltd./OU=RD/CN=Server
Test/Email=server@test]/RootPOA/AccountPOA/ /Audit/d1

...
```

Object domain mapping configuration (ODM.cnf)

7.7 Delegation

7.7.1 Overview

Delegation means that, if an initiator invokes an intermediate object, then this intermediate can, in turn, invoke a target on behalf of the initiator. To indicate the initiator's identity to the target, the intermediate needs to pass some or all of the initiator's credentials (and possibly some of its own credentials) to the target. This delegation process can be repeated several times, so that the initiator's credentials are transferred to the final target through a chain of invocations.

The CORBA security services architecture specifies delegation support of credentials for both security-unaware and security-aware applications. For a security-unaware intermediate object, the specified delegation mode is automatically enforced by the CORBA security services (see Figure 7.1). Whenever the intermediate object invokes a target on behalf of the initiator, the CORBA security services query the intermediate's delegation policy to find out whether the delegated credentials of the initiator or the own credentials of the intermediate (or a combination of both) should become the credentials to be used for the next invocation. If necessary, the intermediate object's principal will also be authenticated by CORBA security services. The

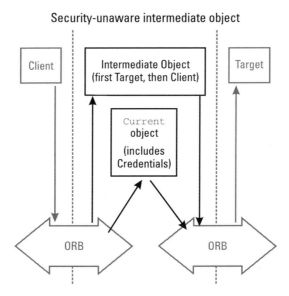

Figure 7.1 Security-unaware delegation.

CORBA security services can retrieve the received initiator's credentials from the intermediate's current object, combine it with the intermediate's own credentials, and set the resulting invocation credentials in Current. These credentials will then automatically be used for invocations from the intermediate to the target. When a target receives an invocation from an intermediate, it needs to ensure that this intermediate is authorized to impersonate the original initiator.

7.7.2   SSL and Delegation

We already explained in Section 6.7 that these described delegation features rely on the capability of underlying security mechanisms to support delegation.

CORBA security services implementations that are only based on SSL as a security mechanism cannot support delegation of Credentials because each SSL endpoint is associated with a specific cryptographic private key. Each private key is linked to a corresponding public key, which in turn is linked to the principal identity in the X.509 identity certificate. The certificate binds the public key to the principal name. This publicly available identity certificate is digitally signed by a certification authority that is trusted by both parties. The SSL authentication process checks whether the

SSL implementation behind the remote socket has access to the private key that corresponds to the public key in the certificate. This mechanism works well as long as no delegation is required. But if an intermediate wants to invoke another target on behalf of the initiator, then it would need to have access to the initiator's private key, which by definition should only be known to the initiator.

7.7.3 CSIv2-SAS Delegation

To overcome this problem, the OMG specified a Security Attribute Service (SAS) as part of the *Common Secure Interoperability* (CSIv2) architecture [3]. It introduces an additional security protocol layer on top of the underlying secure transport mechanism that provides client authentication, delegation, and privilege token functionality. The SAS protocol is modeled after the *Generic Security Service API* (GSSAPI) token exchange paradigm [4] and exchanges its protocol elements in the GIOP service context.

In essence, the SAS protocol allows tokens to be exchanged across a secure underlying transport. The X.509 identity tokens exchanged at the attribute layer allow an intermediate to act on behalf of (i.e., impersonate) some identity other than its own. To accept such a delegated identity, the target either has to trust the intermediate directly or base its trust on a proxy rule certificate (called *authorization token* [5]) that has been signed either by the initiator or a trusted privilege authority. Such a proxy certificate specifies whether the intermediate is authorized to act on behalf of the initiator or not. More details on the CSIv2-SAS delegation protocol can be found in Section 6.7.

Although CSIv2 is security related, it is not specified within the CORBA security services specification but as part of CORBA v2.4. At the time of writing, the relatively new CSIv2 protocol is hardly used in practice, but it is likely that it will be widely used in the near future, first because it runs on top of the extremely widely-used SSL protocol, and second because it is supported by both CORBA and EJB and, thus, enables secure interoperability between both technologies. Currently, no implementation of CSIv2 is available for the standard MICO or MICOSec distribution, but it is anticipated that one will become available in the near future.

7.8 Implementation Overview and Conformance

As we have discussed, MICOSec can secure applications that do not contain any security-related code. To achieve that, various functional components have to be configured with a number of command line arguments and configuration files. As far as security-unaware functionality is concerned, level 1 conformance is a strict subset of level 2 conformance, and so this section only covers the richer level 2 conformance requirements.

The parameters for SSL, which takes care of principal authentication and security association establishment (including mutual authentication and replay/reorder protection), can be supplied as command line arguments. As a result, no modifications to the application source code are required. Due to the fact that SSL is a secure transport layer that replaces TCP/IP as a network protocol, it is, as such, not considered part of the MICOSec security services. Instead, it is integrated into the MICO ORB as an alternative transport layer. For the administrator, however, it appears to be part of MICOSec, because the format of these MICO-specific command line arguments is the same as for other MICOSec security features.

The ODM table is supplied to MICOSec as a command line argument that specifies a mapping configuration file. The file content specifies which X.509 identities, POA names, and object identifiers should be mapped onto which domain names. In addition, it is necessary to manually create the corresponding POA hierarchy from within the target application.

Access control is also configured with command line arguments that name two configuration files. One file associates required access rights with targets, while the other file grants corresponding access rights to the principals. This way, no modifications to the application code are necessary to configure and enforce access control.

Security auditing can also be done without any changes to the source code. The command line parameters specify the audit filter configuration file to use, the type of audit storage, and a storage-dependent parameter (e.g., a file name). Again, filters can be defined on a per-domain basis, using the domain names supplied by the ODM.

As already illustrated in Section 6.7, delegation cannot be supported without an additional CSIv2 protocol layer that supports the use of delegation tokens. With respect to security-unaware applications, it is easy to implement CSIv2 in such a way that the application source code remains unmodified. However, it is most likely that this will involve additional command line arguments to specify delegation tokens.

Most of MICOSec's security enforcement is implemented inside interceptors, which are called by the ORB at several points in the invocation path. An interceptor is a routine that is called by the ORB during the processing of requests. On the target side, it allows access to the GIOP message and request before and after a servant is called.

There are two different types of interceptors in CORBA:

- *Message level interceptors* that offer access to the GIOP message octet stream;
- *Request level interceptors* that handle marshaled GIOP requests.

In MICO, the best place for access control and auditing is the request level interceptor `Interceptor::after_unmarshal` because it provides access to the maximum amount of useful information, such as the operation to be invoked. Message level interceptors are not used by MICOSec since all message protection is provided by the underlying SSL transport layer.

These interceptors can also be used by programmers to add custom-security functionality below the application layer (i.e., without any modifications to the application source code). For example, it is possible to obtain information from the request and security attributes. However, interceptor programming is highly ORB- and version-specific and, thus, will not be covered in detail in this book.

Since this chapter is concerned with CORBA security for security-unaware applications, it describes how the access control features can be configured solely with command line arguments and configuration files (i.e., without any modifications to the application source code). But it is also possible to load and bootstrap configuration files from within the application. Configuration files can be loaded and set up using the MICOSec specific operation `loadConfigFile` on the corresponding security-related objects (e.g., `ODMFactory`, `RequiredRights`). In the same way, the operation `saveConfigFile` can be used to store a created or modified configuration back into a configuration file. Although this makes applications security-aware in CORBA security conformance terms, it is just another way of bootstrapping MICOSec's security-unaware functionality.

7.9 Summary

This chapter covers the use of MICOSec for applications that should remain security-unaware (i.e., the application does not contain any security-related code). As far as security-unaware functionality is concerned, level 1 conformance is a strict subset of level 2. Therefore, this chapter only needs to deal with the richer level 2 conformance requirements to cover both levels. It describes the command line arguments to set up SSL parameters and X.509 identity certificates for principal authentication and security association establishment, as well as configuration files for object domain mapping, access control policies, principal rights, and audit filters.

The standard MICO SSL command line arguments are used to configure MICOSec's principal authentication and security association establishment. This way, certificates and cryptographic keys can be supplied to MICOSec without any active involvement of the application.

MICOSec's ODM functionality also uses a command line argument that points to the ODM configuration file, which contains all the mappings from X.509 identities and POA names (and, optionally, object identifiers) to hierarchical domain names. However, to make this work, the application needs to create a POA hierarchy that matches the ODM domain hierarchy.

The access control and audit features also work for security-unaware applications, again bootstrapped with command line arguments, which point to configuration files that specify required access rights, granted access rights, and audit filters. Both access control and audit policies can use the domain names supplied by the ODM to describe target objects. In addition, the auditing component requires extra parameters to select one of three different audit storages: the standard Unix logging mechanism `syslog`, a flat file, and an SQL database (`PostgreSQL`).

Delegation of `credentials` cannot be supported by CORBA security services implementations that only use SSL transport layer security contexts to set up ORB layer security associations. On the SSL layer, an endpoint is identified by a public key certificate that corresponds to the private key used to encrypt the channel between two endpoints. Now, if an intermediate target node wants to take on the initiator's identity (i.e., impersonate the initiator) when it invokes another target, then it would need to have access to the initiator's private key. However, the trustworthiness of the whole process relies critically on the fact that private keys are never revealed, so that the use of the private key proves the identity associated with the key. To enable delegation, the OMG specified a Security Attribute Service (SAS) as part of the Common Secure Interoperability (CSIv2) architecture [3]. The SAS resides on top of the underlying secure transport mechanism and provides an extra protocol layer that can be used to transfer tokens. This allows initiators to grant the right to be an intermediate to other identities and enables targets to verify if an intermediate node has been endorsed by the initiator.

7.10 Further Reading

The CORBA security services specification [1] only standardizes which security features have to be supported at the ORB layer for security-unaware applications, but it does not specify any use guidelines. As a result, the command line arguments and configuration files covered in this chapter are MICOSec specific. Therefore, they are not described anywhere else in the literature, apart from the *MICOSec User's Guide* [6], which does not go beyond what has been described in this chapter.

References

[1]   OMG, *CORBA Security Services Specification*, Version 1.8 (Draft Adopted Revision), 2000.

[2]   OMG, *CORBA Security Domain Membership Management Service*, Final Submission, July 2001.

[3]   OMG, *Common Secure Interoperability V2 Specification*, March 2001.

[4]   Linn, J., *Generic Security Service Application Program Interface Version 2, Update 1*, (IETF RFC 2743), 2000.

[5]   Farrell, S., and R. Housley, *An Internet Attribute Certificate Profile for Authorization*, (IETF ID PKIXAC), 2000.

[6]   Schreiner, R., and U. Lang, *MICOSec User's Guide*, ObjectSecurity Ltd., 2000, www.micosec.org.

List of Acronyms

ACL   access control list

API   application programmers interface

BOA   basic object adapter

BS   British standard

CA   certification authority

CORBA   Common Object Request Broker Architecture

CORBASec   CORBA Security Services

CSI   Common Secure Interoperability

DBMS   database management system

DCE   distributed computing environment

DCE-CIOP   DCE Common Inter-ORB-Protocol

DII   dynamic invocation interface

DSI   dynamic skeleton interface

DTCB   distributed TCB

EJB   Enterprise Java Beans

ESIOP   environment-specific inter-ORB-protocol

GIOP   General Inter-ORB-Protocol

GMITS   *Guidelines for the Management of IT Security*

GNU   GNU's Not Unix

GPL   GNU General Public License

GSS-API   Generic Security Service API

HTTP   Hypertext Transfer Protocol

HTTPS   Hypertext Transfer Protocol with SSL

ID   identifier

IDL   Interface Definition Language

IDS   intrusion detection system

IEC   International Engineering Consortium

IFIP   International Federation for Information Processing

IIOP   Internet-Inter-ORB-Protocol

IOR   interoperable object reference

IP   Internet Protocol

IR   interface repository

ISO International Standardization Organization

IT information technology

ITSEC IT Security Evaluation Criteria

LAN local area network

LGPL GNU Library General Public License

MDI most derived interface

MICO MICO is CORBA

MICOSec MICO Security Services

NR nonrepudiation

OA object adapter

OBV objects by value

ODM object domain mapper

OMA Object Management Architecture

OMG Object Management Group

ORB Object Request Broker

OS operating system

OSI Open Source Initiative

OTS object transaction service

OU organizational unit

PC personal computer

PKI   public key infrastructure

POA   portable object adapter

POS   persistent object service

QoP   quality of protection

RBAC   role-based access control

RFI   request for information

RFP   request for proposal

RPC   remote procedure call

SAS   Security Attribute Service

SDMM   Security Domain Membership Management Service

SECIOP   Secure-Inter-ORB-Protocol

SESAME   Secure European System for Applications in a Multivendor Environment

SOCKS   Socket Server

SPKM   Simple Public-Key GSS-API Mechanism

SQL   Simple Query Language

SSL   Secure Sockets Layer

SSLIOP   SSL-Inter-ORB-Protocol

STL   Standard Template Library

TCB   trusted computing base

TCP   Transport Control Protocol

TCSEC *Trusted Computer System Evaluation Criteria* (The *Orange Book*)

TIO time interval object

TLS Transport Layer Security

TP transaction processing

UID user identifier

UTO universal time object

VPN virtual private network

WAN wide area network

WG working group

About the Authors

Ulrich Lang received his M.Sc. in information security from the University of London in 1997, after studying computer science with management at the Ludwig-Maximilians-Universität in Munich and at the Royal Holloway College (University of London). Following his graduation, Mr. Lang worked as an independent security consultant on various CORBA-based banking projects. In April 1998, he joined the Security Group at the Computer Laboratory of the University of Cambridge, where he is an active researcher in middleware security. In addition, he is a founder and research director of ObjectSecurity Ltd.

Rudolf Schreiner received his Dipl.-Phys. (physics and astronomy) from Ludwig-Maximilians-Universität in Munich in 1993. After graduation, he worked as a freelance programmer and consultant on various computer security projects. In 2000, Mr. Schreiner became one of the founders and chief technology officer of ObjectSecurity Ltd., a consultancy specializing in distributed systems security.

Index

Multimedia Database Management Systems, Guojun Lu

Practical Guide to Software Quality Management, John W. Horch

Practical Process Simulation Using Object-Oriented Techniques and C++, José Garrido

Secure Messaging with PGP and S/MIME, Rolf Oppliger

Software Fault Tolerance Techniques and Implementation, Laura L. Pullum

Software Verification and Validation for Practitioners and Managers, Second Edition, Steven R. Rakitin

Strategic Software Production with Domain-Oriented Reuse, Paolo Predonzani, Giancarlo Succi, and Tullio Vernazza

Systems Modeling for Business Process Improvement, David Bustard, Peter Kawalek, and Mark Norris, editors

User-Centered Information Design for Improved Software Usability, Pradeep Henry

Workflow Modeling: Tools for Process Improvement and Application Development, Alec Sharp and Patrick McDermott

For further information on these and other Artech House titles, including previously considered out-of-print books now available through our In-Print-Forever® (IPF®) program, contact:

Artech House	Artech House
685 Canton Street	46 Gillingham Street
Norwood, MA 02062	London SW1V 1AH UK
Phone: 781-769-9750	Phone: +44 (0)20 7596-8750
Fax: 781-769-6334	Fax: +44 (0)20 7630-0166
e-mail: artech@artechhouse.com	e-mail: artech-uk@artechhouse.com

Find us on the World Wide Web at:
www.artechhouse.com